Blood Groups of Primates: Theory, Practice, Evolutionary Meaning

MONOGRAPHS IN PRIMATOLOGY

Editorial Board

Volume 1: Child Abuse: The Nonhuman Primate Data
Martin Reite and Nancy G. Caine, *Editors*

Volume 2: Viral and Immunological Diseases in Nonhuman Primates
S.S. Kalter, *Editor*

Volume 3: Blood Groups of Primates: Theory, Practice, Evolutionary Meaning
Wladyslaw W. Socha and Jacques Ruffié

Blood Groups of Primates: Theory, Practice, Evolutionary Meaning

Wladyslaw W. Socha

Research Professor of Forensic Medicine
Director, Primate Blood Group Reference Laboratory at the Laboratory for
Experimental Medicine and Surgery in Primates (LEMSIP) of the New York University
School of Medicine, New York
Associate Director, WHO Collaborating Centre of Haemotology
of Primate Animals

Jacques Ruffié

Professeur au Collège de France
Director, Laboratory of Physical Anthropology, Collège de France, Paris
Professor, New York University School of Medicine, New York

ALAN R. LISS, INC., NEW YORK

Address all Inquiries to the Publisher
Alan R. Liss, Inc., 150 Fifth Avenue, New York, NY 10011

Library of Congress Cataloging in Publication Data

Socha, Wladyslaw W., 1926–
  Blood groups of primates.

  (Monographs in primatology; v. 3)
  Bibliography: p.
  Includes index.
  1. Blood groups in animals. 2. Primates— Physiology.
3. Primates—Evolution. I. Ruffié, Jacques. II. Title.
III. Series.
QP98.S6 1983   599.8'04116   83-14990
ISBN 0-8451-3402-7

TO JAN MOOR-JANKOWSKI

Contents

List of Figures xi

List of Tables xiii

Introduction xvii

1 The Place of Primates in the Animal Kingdom 1

Origins 1
Characteristics of Nonhuman Primates 2
The Importance of Learned Behavior for the Primates 8

2 Taxonomy of Living Primates 11

Prosimii 11
Anthropoidea 14
  Platyrrhini 14
    Callitrichidae 14
    Cebidae 14
  Catarrhini 15
    Cercopithecoidea 15
    Hominoidea 20

3 Monkeys and Man 23

Monkeys in History 23
From Nonhuman to Human Primates 25
Utilization of Primates in Medicine 26

4 History of the Discovery of Blood Groups in Monkeys 31

Discovery of the First Human Blood Groups 31
Immunological Polymorphism of Living Organisms 32
Discovery of Blood Groups in Domestic Animals 35
Discovery of Blood Groups in Nonhuman Primates 36
Problems of Nomenclature 38

5 The A-B-O System 39

The Nature of the Antigens of the A-B-O System 39
The Genetic Model of the A-B-O System 39
The Subgroups of A: A_1 and A_2 45
Variations of A-B-O Polymorphism in Nonhuman Primates 51

6 The M-N System 53

The Genetic and Phylogenic Complexity of the M-N System   53
The M-N Series   54
    Review of Discoveries   54
    The M-N Blood Group System in Man and Anthropoid Apes   56
    The Phylogenic Evolution of the Chromosome Segment Affecting the M-N
    System   57
    The Genetic Model of the M-N System According to Wiener et al.   59
    Mutations Specific to Each Species   63
    The Current Genetic Model of the M-N System   65
The Henshaw (He) and Hunter (Hu) Antigens   66
Antigens of the Miltenberger Series   67
Antigens of the S/s/U Series   68
Antigens of the $V^c/A^c/B^c/D^c$ Series   69
Antigen W^c   70
Phylogeny of the M-N Blood Group System   71

7 The Rhesus System 75

History of the Discovery of the Rh System   75
The Rhesus System in Man   76
The Rhesus System in Nonhuman Primates   78
    Rh Factors Recognized by Reagents of Human Origin   78
    Rh Factors Recognized by Reagents of Simian Origin:
    The $C^c/c^c/E^c/F^c$ Series   81
    Genetic Model of the R-C-E-F System of Chimpanzee   84
    The R-C-E-F System in Other Nonhuman Primate Species   88
The Rhesus System in Evolution   89

8 Blood Systems Specific to Cercopithecoidea (Old World Monkeys) 91

The Graded D^{rh} Blood Group System of Macaques   91
The Graded B^p System of Baboons   94
Relationship of the Graded Blood Group Systems of Old World Monkeys to
Those of Anthropoid Apes and Man   96

9 Methodology of Blood Grouping in Nonhuman Primates 99

Development of Testing Techniques   99
Collection and Shipment of Specimens for Blood Grouping   106
Tests With Reagents Originally Prepared for Typing Human Blood   109
    Tests for A-B-O Blood Groups   109
    M-N Typing   120
    Rh-Hr Typing   120
    Lewis Typing   128
    Testing for I and i Blood Factors   129
    Tests for Homologues of Other Human Blood Groups   132
Tests With Reagents Produced by Immunization of Primate Animals   132
    Immunization   133
    Testing of Antisera   138

Production of Antiglobulin Sera. 139
Immunization Protocol 141
Plasmapheresis 143
Standardization of Typing Reagents. 143
Typing Reagents 153
Monoclonal Antibodies 158

10 Spontaneously Occurring Agglutinins in Primate Sera   163

Discovery. 163
Occurrence in Various Species 165
Classification and Significance 173

11 Practical Applications of Blood Group Studies in Nonhuman Primates   179

Serological Maternofetal Incompatibility 179
Homologous Transfusion 188
Heterologous Transfusion. 201
Transplantation 203
Blood Groups as Genetic Markers; Paternity Investigations. 206
Seroprimatology 212
Primate Immune Sera as Typing Reagents for Human Red Cells. 219

12 Prospects Offered by the Study of Blood Groups of Nonhuman Primates   223

Comparative Study of Blood Groups in Populations and Evolutionary
Mechanisms. 223
Types of Blood Factors in Living Species 224
Paleosequences and Neosequences 227
Genetic Polymorphism and Speciation 228
Causes of the Maintenance of Polymorphism and Its Variations in Populations of
Nonhuman Primates 230
Role of Sequential Redundancies in the Diversifying Evolution of Species 236

Bibliography   239

Index   255

List of Figures

1-1 Distribution of nonhuman primates within the ecologic zones
     of the world   3
1-2 World distribution of nonhuman primate families.   5
2-1 Taxonomy of living primates 12
2-2 Geographical range of baboons 16
5-1 Evolution of the A-B-O groups in primates. 41
5-2 Successive appearance of the genes of the A-B-O blood group system
     in the course of the immunologic evolution of primates 43
5-3 Distribution of A and H activities over A and H variant glycolipids
     from group A_1 and A_2 erythrocytes 49
5-4 Schematic representation of A and H glycolipid status in group A_1 and
     A_2 erythrocytes 50
6-1 Hypothetical comparison of the chromosomal segment of the M-N
     loci of man and of chimpanzee 73
8-1 Two immunological stocks, one corresponding to the baboon phylum,
     and the other to the rhesus phylum, may have emerged from a
     common ancestor that had antigenic structures, shared by B_3 of
     baboon and D_3 of macaques 96
8-2 Diagram showing how the "step-like" mutations that invoked a
     common ancestral factor led to the formation of the graded subtypes,
     either by acquisition of new factors (toward B_4 and D_4 types) or by
     loss of certain patterns (B_1 and D_1 types). 97
11-1 Survival of ^{51}Cr-labeled red cells cross-tranfused into recipient animal
     (rhesus monkey or baboon) 198
11-2 Survival of the ^{51}Cr-labeled red cells cross-transfused into recipient
     pairs of baboons 199

List of Tables

2-1   Classification of primates proposed by Hoffstetter   22

4-1   Constitution of the four basic blood groups of the A-B-O system   32

5-1   A-B-O blood groups of apes and monkeys   46

6-1   The M-N system as conceived at the time of its discovery   55

6-2   The M and N blood group factors in anthropoid apes   57

6-3   Relationships among M-N genotypes and phenotypes according to
Wiener's hypothesis   59

6-4   Genetics of the M-N system of man (and gibbon) according to Wiener's
hypothesis   60

6-5   Genetics of the M-N system of chimpanzee according to Wiener's
hypothesis   61

6-6   Genetics of the M-N system of gorilla according to Wiener's
hypothesis   62

6-7   Genetics of the M-N system of orangutan according to Wiener's
hypothesis   64

6-8   Alleles of the V-A-B-D system and their presumed action   70

6-9   Serology and genetics of the chimpanzee V-A-B-D blood group
system   71

6-10   Inhibition tests with glycoprotein preparations obtained from the red
cells of chimpanzees of various V-A-B-D and W^c types   72

7-1   Comparison of two nomenclatures for the Rh blood group system . . .   76

7-2   Results of comparative titration and absorption experiments with
chimpanzee anti-R^c and human anti-Rh_0 sera   82

7-3   Subdivision of the type c^c by means of anti-c^c and anti-c_i^c reagents . . .   85

7-4   Postulated alleles of the R-C-E-F system and their products   85

7-5   Serology and genetics of the chimpanzee R-C-E-F blood group system   87

8-1   Anti-D^{rh} typing reagents   92

8-2   Serological definition of blood groups of the D^{rh} graded system and
their distribution among various species of Old World monkeys   93

8-3   Genetics of the D^{rh} system as observed in a sample of unrelated pig-
tailed macaques (*Macaca nemestrina*)   95

9-1   Comparative titrations of an unabsorbed anti-B serum of human origin
against primate and human red cells suspended in isotonic saline
solution   111

9-2   Representative titrations of the normal sera of apes against human red
cells of various A-B-O groups   113

9-3   Protocol of tests for A-B-O blood groups in anthropoid apes   114

9-4   Example of a protocol of the A-B-O blood grouping tests (quantitative saliva inhibition tests and reverse serum tests) in Old and New World monkeys   116

9-5   Representative results demonstrating the specificities A, B, and H on red cells and in saliva, and anti-A or anti-B agglutinins in sera of 10 mantled howler monkeys (*Alouatta palliata*) among a series of 52 monkeys, with comparative reactions for man, marmoset, rabbit, rhesus, baboon, and gorilla   118

9-6   Demonstration by absorption of the differences among the B agglutinogens on the red cells of man, howler monkeys, marmosets, and rabbits   121

9-7   Results of tests on red cells of gorillas, orangutans, and other simians, with anti-M, anti-M^e, anti-He, anti-N^V reagents   122

9-8   Fractionation by absorption of the anti-M reagent (M_1) prepared from an immune rabbit serum for human red cells   124

9-9   Results of tests with Rh antisera on blood specimens of 14 chimpanzees   126

9-10 Results of tests with Hr antisera on blood specimens of 14 chimpanzees.   127

9-11 Comparison of orangutan, chimpanzee, and gibbon red cells in tests with human anti-hr' serum.   128

9-12 Results of comparative titration and absorption experiments with human anti-Rh_0 serum   129

9-13 Results of inhibition tests for substances H and Le in the saliva of primates.   130

9-14 The blood factors I and i in infrahuman species   134

9-15 Comparison of results of titration by the saline agglutination, antiglobulin, ficinated red cell, and dextran methods.   140

9-16 Protocol of immunization of a chimpanzee   142

9-17 Protocol of immunization of a rhesus monkey   144

9-18 Comparison of titrations of serum of isoimmunized chimpanzee by three different methods.   145

9-19 Results of titration of the serum of an isoimmunized chimpanzee and the presumed specificities of the antibodies detected   146

9-20 Titration of serum of chimpanzee Leo, No. 335, before and after multiple absorptions with the red cells of D^c-positive B^c-negative type of chimpanzee Gabriel, No. 19 (composite table of findings)   148

9-21 Results of test with anti-O^c + P^c and anti-P^c reagents carried out by enzyme-treated red cell method and blood samples of 45 chimpanzees   149

9-22 Contingency tests for associations among specificities L^c (R^c), C^c, c^c and P^c (c_1^c).   150

9-23 Elimination of prozone by inactivation of isoimmune rhesus monkey serum No. 228   152

9-24 Cross-reactions of two anti-A^P isoimmune baboon sera, with red cells of macaques and gelada monkeys (saline agglutination method)   152

9-25   Comparison of the distribution of red cell specificities, defined by pairs of rhesus isoimmune sera of supposedly identical specificities, in six species of macaques   153

9-26   Rhesus isoimmune sera used for blood grouping tests in macaques   . . .   154

9-27   Baboon blood grouping reagents   156

9-28   Standard reagents used for chimpanzee blood grouping   158

10-1   Results of cross-tests by the antiglobulin method on the sera and red cells of ten baboons   166

10-2   Results of titration of a type-specific antibody discovered in an apparently normal female baboon   167

10-3   Comparative results obtained with the red cells of hamadryas baboons in tests with antibody-containing normal baboon sera and with standard anti-B^P isoimmune reagent.   168

10-4   Results of cross-matching tests by the saline agglutination method on the sera and red cells of 18 nonimmunized crab-eating macaques (*Macaca fascicularis*)   169

10-5   Comparison of the reactions of two naturally occurring isoagglutinins in the sera of bonnet macaques   170

10-6   Results of cross-matching tests by the antiglobulin method on the sera and red cells of 11 nonimmunized bonnet macaques (*Macaca radiata*) . .   171

10-7   Comparison of the reactions of naturally occurring agglutinins in the sera of pig-tailed macaques with enzyme-treated red cells of macaques   172

10-8   Titration of autoantibodies in baboon serum against ficinated red cells, showing effect of temperature on the reactions.   173

11-1   Results of studies on three pregnancies of crab-eating macaques isoimmunized with the breeding mate's red cells.   182

11-2   Titers of antibodies in the serum of a pregnant orangutan female against red cells of the breeding mate.   185

11-3   Blood groups of the members of an orangutan family with cases of erythroblastosis   187

11-4   Pairing of rhesus monkeys and baboons for cross-transfusion   189

11-5   Antibody titer and survival of transfused red cells following multiple transfusions of isologous blood in rhesus monkeys.   192

11-6   Antibody titer and survival of transfused red cells following multiple transfusions of isologous blood in baboons   194

11-7   Serological findings and posttransfusion recovery of human type O RBCs in nonhuman primates.   204

11-8   Blood components as possible genetic markers in chimpanzee   208

11-9   Exclusion probabilities estimated from 12 polymorphic red cell proteins and two blood group systems in *Macaca nemestrina* of Malaysia   209

11-10   Two examples of the use of blood groups for solving problems of parentage in chimpanzees   211

11-11   Example of the use of blood grouping tests for solving problems of parentage in chimpanzees   213

11-12   Distribution of the V-A-B-D blood groups in common and pygmy
        chimpanzees.   214
11-13   Distribution of the R-C-E-F blood types in common and pygmy
        chimpanzees.   216
11-14   Distributions of red cell specificities (defined by rhesus isoimmune sera)
        in six species of macaques.   218
11-15   Blood grouping tests in an egg transfer experiment   220
 12-1   Measurements of the polymorphisms in some species of nonhuman
        primates.   232

Introduction

The discovery and identification of the blood groups of man is one of the most important achievements of modern biology. At the beginning of this century, Karl Landsteiner described the A-B-O system, the knowledge of which enabled blood transfusion to become current practice, thus opening the way to the use of modern surgical methods. In 1927, he discovered, with Levine, the M-N and the P systems; and in 1940, with A.S. Wiener, the Rhesus system, to which were added, after World War II, the Lutheran, Kell, Lewis, Duffy, Kidd, Diego, Auberger, Xg^a, and other groups.

The blood groups revealed, for the first time, our genetic polymorphism, which until then had not been suspected. Because of their frequency and ease of identification, the blood groups permitted us to apply to our species mathematical models of population genetics. The knowledge of the heredity of man, hampered by the impossibility of experimental breeding, was thus greatly enlarged. To these red cell blood groups many other systems, either immunological (such as the HLA system, situated on the white cells and the body tissues) or enzymatic, were added. They constitute "markers," which have enabled us to label the chromosomes of the human karyotype and to draw up the first charts of our genetic patrimony.

Meanwhile, the study of the distribution of the blood factors in the populations of the world was changing traditional biological anthropology. Until then, our species had been thought of as divided into races, each of which consisted of individuals bearing the same characteristics and which were always identical to a basic model, the *holotype*. The analysis of blood groups has put an end to this typological view of humanity, by demonstrating that no one race has the prerogative of a single blood group and that all the groups are found in all races, only their frequencies varying from one region to another. Today, humanity no longer appears as a collection of separate races but rather as a series of genetically polymorphous populations, within which individuals have more of an opportunity to mate among themselves than with individuals of neighboring populations. Nevertheless, no population is really closed; there are almost always a number of interpopulational exchanges thanks to which humanity constitutes a single straight pool of intercommunicating genes. This definition of an individual and a population has given rise to a new science, which was named *hemotypology* [172]. After an atlas had been drawn up of the worldwide distribution of blood types (or hemotypes) observed in various regions, the role of chance (genetic drift), which operates especially among strictly isolated populations of small size, was evoked. But there also seemed to exist differences in selective pressures, which could have favored one gene here and its allele elsewhere; for example, natural selection due to maternofetal immunological incompatibility seemed to play a role in the formation of current blood group polymorphisms [194]. However, exogenous pathological factors (parasitic, viral, or bacterial endemics) must play a substantial role in the selective survival of individuals [147]. Variations in gene frequencies that occur in space must be the result of migrations, which have always been an attribute of man. In order to detect these variations, it was necessary to demonstrate correlations between the distribution of cultural (or linguistic) traits and the distribution of hemotypes [6,20,21]. Thus, we can evaluate the influence of prehistoric events (the spread of the "neolithic revolution," for instance) and historical events (invasions, great epidemics) on the biological structure of human populations.

But there is a third approach to the study of blood factors: we can consider them on a phylogenic level and observe how they were transformed during the different evolutionary stages that eventually led to man. None of the human attributes, especially the most complex, appeared abruptly *de novo* with the arrival of our first ancestors. They must have been prepared, perhaps long before, in the species that preceded man, of which some still living descendants carry the traces.

Let us examine the example of our brain, which is a remarkably complicated organ. In the nonhuman primates it exists in a simpler form and as a more and more elementary structure in reptiles, amphibians, and fish. We cannot fully understand the brain of man unless we consider what that organ was in the species that preceded us, and how, little by little, the different parts of which it is composed today took their place.

The same approach must be used to grasp the structure of the immunological systems. Systems as complex as the Rhesus or the M-N undoubtedly existed, at least in a rough form, in the infrahuman stages, especially in the simian primates and, above all, in the great anthropoid apes. All these species must have come from a common branch during the Cenozoic era. In the middle of this period, toward the beginning of the Miocene, the human lineage separated itself from the branch that produced the great anthropoid apes. The species still living (the Asian orangutan and, above all, the African gorilla and chimpanzee) have preserved some traces of the ancestral state. Those ancestral traits may help us to clarify the manner in which, during the last 20 or 30 million years, the immunological systems of human red cells were formed.

Unfortunately, the study of immunological systems is much more difficult to carry out with monkeys than with human beings. Human populations are numerous, easy of access, and relatively docile, at least when it comes to medical problems. Moreover, the development of a network of transfusion services throughout the world and the common application of systematic serological examination of pregnant or delivering women during the last 30 years make immunological research in humans relatively easy.

It is very different with the nonhuman primates, whose numbers are infinitely fewer and who form colonies difficult to control in the wild. In addition, the techniques of immunology depend largely on alloimmunization, either spontaneous (pregnancy) or induced (blood transfusion, injection of blood). With monkeys, such conditions are possible and controllable only by the breeding in captivity of large numbers of animals, which is always troublesome and difficult.

From the beginning, it was felt that human immunological polymorphism ought to be found in monkeys also. This was confirmed as early as 1925, by Landsteiner and Miller, who found that chimpanzees could be group A or group O. Later on, Landsteiner and Wiener attempted to detect other antigens common to man and other primates. In 1940, while pursuing this line of investigation, they made a series of observations that were to lead to the definition of the Rhesus system. A.S. Wiener, later joined by J. Moor-Jankowski, pursued these investigations, concentrating on the problem of the phyletic origin of the blood factors that are found on both monkey and human erythrocytes but have evolved in a different manner. They were remarkably successful in describing a series of simiam blood groups characteristic of certain primate species and immunologically equivalent to the human blood groups.

The study of the blood groups of nonhuman primates—particularly, comparative immunological investigations—was greatly facilitated by the creation of the Laboratory for Experimental Medicine and Surgery in Primates (LEM-SIP) at the New York University School of Medicine through the initiative of Jan Moor-Jankowski. This laboratory of sophisticated technology, originally devoted to erythrocyte immunology, was organized in connection with an important colony of primate animals, which provided priceless materials for the study of various species of monkeys and anthropoid apes.

The conjunction of these two factors was of utmost importance for the accumulation of the experience and data without which this book would have not been possible. It is our pleasure, therefore, to dedicate this work to our friend Jan Moor-Jankowski, as the expression of our appreciation of his role in the progress of this chapter of primatology.

Primatology has been called upon to play a greater and greater role in biomedical research. During the last century, physiologists, followers of

Claude Bernard, were largely involved in describing macrofunctions. Their research could be done with dogs or rats: all mammals showed comparable reactions at this level. The regulation of glycemia or hematopoiesis, for example, is the same *grosso modo* for all mammals, and we can understand what happens in man by studying other, far removed species. But the situation is very different when we leave the macrofunction stage to enter the field of microfunctions, cellular, immunological, and molecular. At this level, differences between species are seen and are all the greater as one deals with species further separated from each other on the phylogenic tree. Also, an increasing number of investigations in human medicine require the use of primates that are the closest to us; our "biological cousins" are the only ones that can provide a model close enough to make possible the solution of many problems of human pathology. Thus, the higher primates represent an irreplaceable source for today's investigator and even more for tomorrow's.

For example, we know that the chimpanzees, the animals in many respects the closest to us, carry 98% of our common genetic information. Their disappearance, like that of the other species of nonhuman primates, would be an irreparable loss. The nonhuman primates, long hunted irresponsibly and unprotected, are becoming rare. Some countries have forbidden their exportation so that their population census may increase. Those countries that follow a laissez-faire policy have almost no more animals. Hence the imperative need to create, everywhere, breeding centers where these species, precious for us and for those who will come after us, can be maintained and protected. It is well known by now that breeding in captivity requires a thorough knowledge of the genetics of the animals in order to prevent the effects of inbreeding. At the current state of the art, the blood groups constitute the most important set of chromosomal markers readily available to breeders of nonhuman primates [195]. This was recognized, for example, by the recent Report of the Ad Hoc Task Force to Develop a National Chimpanzee Breeding Plan, which recommended that blood grouping tests be routinely carried out on all chimpanzees in the United States, as an integral element of the rational breeding of these endangered apes in captivity [161a].

Most biomedical research based on the use of nonhuman primates presupposes a thorough knowledge of the immune constitution of the experimental animals. Therefore, in writing this book we wished to prepare a balance sheet of what is known at present about the immunology of the red cells of primate animals as a guide to the better understanding of the human blood groups and as an aid to clarification of some immunological processes associated with the differentiation of species. Several chapters are devoted to techniques that are as yet little known in immunohematology laboratories but will have to be used extensively in the future.

We wish to thank our friend Arthur E. Mourant and Drs. Charles Salmon, Philippe Rouger, and Jean-Pierre Cartron for suggestions and recommendations offered at various stages of preparation of the manuscript. We also thank Ms. Mollie Saltzman, who has reviewed the English translation of part of the text. Finally, we thank Ortho Diagnostic Systems, Inc., Raritan, New Jersey, for its generous donation, which made completion of this work possible.

Wladyslaw W. Socha
Jacques Ruffié

Blood Groups of Primates, pages 1–10
© 1983 Alan R. Liss, Inc., 150 Fifth Avenue, New York, NY 10011

1

The Place of Primates in the Animal Kingdom

Origins   1
Characteristics of Nonhuman Primates   2
The Importance of Learned Behavior for the Primates   8

ORIGINS

The primates are one of the most recently emerged mammalian orders. Their ancestors evolved at the end of the Mesozoic (the genus *Purgatorius* of the Upper Cretaceous) from small nonspecialized insectivores, of which the tree shrews (tupaia), living on the east coast of India in Southeast Asia and Indonesia, give us some idea. This group broke up into several branches at the beginning of the Cenozoic (paleocene epoch) 80–100 million years ago. Later, each lineage dichotomized in its turn. The first known primates are the prosimians, or lower primates, which are about the size of a squirrel and have long tails and four five-fingered limbs with opposable thumbs, a fundamental primate trait whose importance will be discussed later. They lead an arboreal life. During the lower Cenozoic they must have spread rapidly throughout the Old and New Worlds. Since the Northern Hemisphere was not yet divided by the Atlantic Ocean, they were able to colonize an enormous territory with numerous ecological niches. The prosimians disappeared little by little. Only a few species remain in relatively well-protected zones where competition is weak (sub-Saharan Africa, Southeast Asia, the southeast coast of India, and Madagascar). Although it is an ancient group, the prosimians already show all the essential primate characteristics (which will be discussed later).

Fairly rapidly, during the Oligocene, a branch split off from this group to form the suborder Anthropoidea, which includes monkeys, apes, and man. During the course of the Oligocene, this suborder split further into several branches. The three major branches are the following:

1. The cynomorphes (or simians), which correspond to the tailed monkeys. At the beginning of the Miocene (25–30 million years ago) they split into two branches that are today geographically isolated: the Platyrrhini, showing

more archaic traits, which inhabit the Americas, and the Catarrhini, more evolved, which inhabit the Old World.

2. The hominids (Hominoidea), which were also divided, at the begining of the Miocene, into two major branches: the Pongidae, or anthropoid apes (great tailless apes), with an Asian branch (orangutan) and an African branch (gorillas and chimpanzees); and the Hominidae, which represent the human lineage marked out by a certain number of stages (Australopithecinae and *Homo habilis, Homo erectus* [Pithecanthropi], *Homo sapiens neanderthalensis,* and *Homo sapiens sapiens*).

3. An intermediate group, found only in Asia, made up of fairly large animals that had no tails (which relates them somewhat to the anthropoid apes) but retained certain traits of the cynomorphes. They correspond to gibbons in the larger sense of the term (the Hylobatidae).

CHARACTERISTICS OF NONHUMAN PRIMATES

All primates inhabit the intertropical zone, including its deserts and its high mountains covered with permanent snow, as shown in Figure 1-1. They prefer savannahs, steppes, and forests.

The primates constitute a relatively homogeneous group that occupies a particular place among mammals. Their principal characteristics may be defined as follows:

1. *Primates show little or no specialization.* The preceding orders display multiple forms of adaptation (for swimming, climbing, flying, digging, etc.). Such is not the case with primates, which do not show any notable traits of adaptation and remain surprisingly homogeneous. This nondifferentiation is exemplified by a set of remarkable traits. Their *dentition* is made up of relatively unspecialized elements (none of their teeth has undergone the adaptive modifications, or has acquired the functional value, of the carnivores' canines, the rodents' incisors, etc.). Their *limbs* have preserved a generalized structure: they retained the clavicle, which is considerably reduced or absent in other mammals. Their pentadactyl *hands* have fine, agile fingers with nails rather than claws, and the pads of the fingers are a sensitive tactile area. Their *muzzles* are shortened.

This absence of specialization is to be found also in most of their physiological functions. Although they are predominantly vegetarian, the primates will eat almost anything. The species living on the ground are voluntarily omnivorous (which corresponds to the most primitive regimen of mammals). Others are sporadically carnivorous.

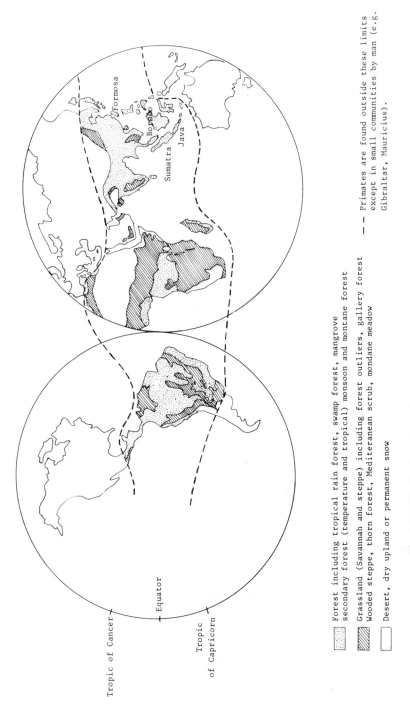

Fig. 1-1. Distribution of nonhuman primates within the ecological zones of the world.

Forest including tropical rain forest, swamp forest, mangrove secondary forest (temperature and tropical) monsoon and montane forest

Grassland (Savannah and steppe) including forest outliers, gallery forest Wooded steppe, thorn forest, Mediteranean scrub, mondane meadow

Desert, dry upland or permanent snow

-- Primates are found outside these limits except in small communities by man (e.g. Gibraltar, Mauricius).

Tropic of Cancer

Equator

Tropic of Capricorn

Formosa

Borneo

Sumatra

Java

Although the primates originated in the tropics, and still show some preference for the warmer regions of the globe, they can stand cold climates and high altitudes (Fig. 1-2). *Macaca mulatta lasiotus* has been seen in China at altitudes up to 4,400 meters and *macaca thibetana* at 3,000 meters. Thus, the primates seem to be able to face up to multiple ecological stiuations without falling into an irreversible and limiting pattern of organic specialization. Because of this, they were available for the ulterior evolution that has led up to *Homo sapiens*.

The only specializing tendency, quite subtile, that can be observed in primates is their adaptation to arboreal life. Tree living represents an advantage for these animals because they have few competitors for their ecological niche. Other than the birds, there are almost no higher vertebrates, and there are few predators. The natural resources are very abundant in this environment for leaf, fruit, and insect eaters. Everything that facilitates arboreal life would probably be retained by natural selection. This doubtless caused the appearance of other characteristics of this group of animals, which are listed below.

2. *Primates display a singular development of sensorimotor faculties,* expressed by the following set of characteristic traits.

Vision is developed to the detriment of the sense of smell. An animal that moves on the ground in a two-dimensional world uses its sense of smell (and also hearing) to hunt or to flee its hunters. Smell is useful to the hunter because the stimulus remains when the quarry is no longer visible, so that its trail can be followed, whereas the animal that depends on its eyes must see the prey or the menacing predator; the stimulus disappears with the object. However, for the animal that jumps from tree to tree in search of food or a mate, sight is more useful than sense of smell. From its tree, the monkey must be able to recognize the fruits and buds that it wants to eat. When it jumps, it must carefully judge distance and the exact position of its landing place. Primates have thus developed stereoscopic vision, their eyes being located on the front of the skull in the same plane, which gives two different views of the same visual field and permits depth perception. In addition, they see colors very well, which enhances their aptitude for recognition.

The primates also have a highly developed sense of balance, necessary for life in the trees; and great precision in their gestures, which are useful in their arboreal life. Most primates are functionally four-handed. Their hands are usually prehensile because the thumb is opposable. They can deftly manipulate the things they find, sometimes transforming them for use in hunting, gathering, etc.

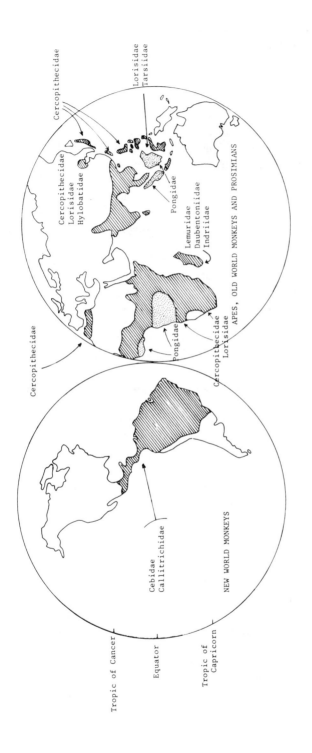

Fig. 1-2. World distribution of nonhuman primate families.

3. *All these sensorimotor faculties are at the service of a well-developed intelligence.* Primates take in information, analyze it, and react with a response appropriate to the given situation. They are able to produce complex strategies, sometimes collectively. This is the result of the development of their brains. Except for the most primitive members of the order, primates have a much larger brain for their body size than the other mammals. This enlargement is found especially in the cortical region, where the higher sensory and motor centers are located. There is surely a link between the enlargement of the cerebral cortex and the acquisition of precise gestures and stereoscopic color vision.

There is a permanent feedback cycle between the brain and the hands. The brain receives, analyzes, and stores data, and the hands execute the brain's orders. The brain then perceives the results of these actions and compares and memorizes. It also may be learning to improve its responses. Selective pressure should strongly favor the development of the link between brain-sensory organs and extremities, which causes a rapid development of the brain, a concomitant increase in psychic capacity, and an increase in the efficiency of the individual animal.

4. *This mechanism should explain the tendency shown by some primates to adopt an upright position.* The gradual acquisition of a semiupright posture was accompanied by a progressive modification of the posterior limbs, which become lower limbs, used almost exclusively for walking, with plantar toes more or less reduced and less prehensile, and a nonopposable hallux. The anterior or upper limbs may be freed from locomotor function and are available for other, more delicate actions. The hand replaces the jaw as an instrument of manipulation. Used only for chewing, the muzzle decreases in size, permitting the enlargement of the brain and cerebral cranium. The tendency to upright posture is very old, but is achieved fully only in man. Undoubtedly, it represented the main force in hominization [178].

5. *The development of their psyches allowed the primates to become social animals.* There are many social mammals; this results from the physiological requirements of their lives. When the young mammals are nursed by the mother for a relatively long time, they establish strong ties to the adult community and especially to the mother. During a long growth period the young animal learns many behavior patterns from the adults. He practices them while playing with the other young in the group, because games simulate, in a danger-free manner, the situations confronted by adults. We now have ample evidence for the value of games in the education of young mammals. The primates exaggerate and develop these tendencies of the lower mammals.

In the order Primates, only the prosimians regularly have multiple births. Starting with the simians, usually only one baby is carried at a time, which allows the mothers to take greater care of their offspring. The perinatal death rate is low compared to that of the lower vertebrates (first are the amphibians and reptiles), which often lose a large number of eggs, tadpoles, or young. Despite a relatively low reproductive level, primates are able to maintain their numbers under natural conditions. However, when man hunts the other primates, the result is often catastrophic for them because of their low reproductive level and long childhood. Man is the only predator who can efficiently conquer the ecological niche of the other primates and who can rapidly destroy them. Because of man, many primate species are today in danger of extinction. It has been suggested that the ancestors of man were responsible for the disappearance of other hominids in the past.

Primate social organization has many forms. Some species lead solitary lives, with a family group limited to mother and baby, but this elementary structure is rare among the primates. Others live as relatively stable monogamous couples, similar to those observed in some birds. The gibbons of southeast Asia follow this pattern of long-lasting families including the mother, the father, the baby, and several immature young. The female has one baby at a time, as do almost all simians. The gestation time is 7 months, and there is a birth every 2 or 3 years. The young gibbon is able to take care of himself at about the age of four. Then he leaves his family group and lives alone until he finds a young female and they form a new couple. The length of the periods of gestation and maturation of the young explains why the gibbon family must remain together so that the baby may be adequately protected and nurtured until he is able to fend for himself. Most monkeys, however, live in larger and more complex groupings of several adults. The most usual is a harem system, which includes a dominant male and several females accompanied by their young, while young bachelor males live on the periphery of the group. Groups including several males are also seen, as well as groups in which males and females mate freely with each other. Whatever the social structure, the hierarchy is rigid. A leader directs the group in its activities and movements in a well-defined territory.

It appears that the social structure is conditioned by ecological imperatives. The harem structure seems to be found in semiarid regions and in the tropical forest, i.e., where the available food is sparse and widely spread. In the territory that it knows well, the group is safer than elsewhere. The resources of the territory are used by the reproductive group—the harem—while the celibate males are left to explore the periphery of the territory, where it is more risky to search for food, which is sometimes very far away. Sometimes, several harems are united in a troop that gathers in the evenings, although each harem group remains together. The hamadryas baboon of Sudan, Ethiopia, and Arabia follows this pattern. The troops gathered at nightfall may contain 200–300 animals in harem groupings of 10–20 members. Each harem is composed of a dominant adult male, four to six females with their offspring, and two or three young adult males (the pages) who live with the group but do not, in principle, mate with the females. When the group travels, the leader's superiority is strengthened: he leads and chooses the direction to be followed, and the others obey. Other males act as sentinels on the outskirts of the group. In case of danger, the males face the aggressor while the females flee with their young. The males run only when the rest of the group is safe. The great apes of Africa, gorillas and chimpanzees, have a similar structure, but the number of animals in the troop is always less.

It might seem that the harem structure, in which only one male is allowed to reproduce, would severely limit the gene pool. In fact, despite the limited risk that it represents, the males on the periphery often profit from the distraction or absence of the leader to couple with a female, who never refuses. In addition, the dominance of one male is not permanent: when the leader is old, an active young male replaces him and goes through a period of intensive reproductive activity. The dethroned leader may live alone, but more often takes up a place on the edge of the group, where he may participate in its defense.

Although a member of a group, the individual primate always maintains a certain degree of freedom. He may take the initiative, though his independence is limited by his group membership. The group imposes a hierarchy and rules but helps and protects as well. Thus, the advantages of a social life outweigh its disadvantages. The solitary animal, integrated into a more or less hierarchical group, profits from the activity of its companions. The baboon in a troop is not limited by what his own eyes can see; he uses the information obtained by each group member and thus, in a way, the sensory organs of his companions. If one baboon spots danger and cries, all the troop members prepare to defend the troop.

THE IMPORTANCE OF LEARNED BEHAVIOR FOR THE PRIMATES

In contrast to insect societies, in which individuals linked by inborn behavior are grouped, and in which each member occupies approximately the same function throughout its lifespan, mammalian societies—especially those of primates—are made up of individuals linked by behavior that is partly inborn but is greatly modified by education.

We have already spoken of the importance of the growth period in the "learning phase," during which a young animal acquires from the adults a behavior system, tests it, and elaborates it. The social insect is a prisoner of genetic programs. The species can only modify them by the processes of natural selection, which are always very slow. Learned behavior, on the other hand, is always more pliant. However, it is also more fragile: when the selective necessities that produced it disappear, a social structure may quickly disintegrate. The *Presbytis entellus* of India used to live in very hierarchical groups in the jungle. Some colonized the temples, where they were protected by man and profited from the abundant food offered to the monks by pilgrims. As they became beggarly and parasitic, the groups lost all cohesion and hierarchal organization.

This fragility does offer an advantage—that of allowing an animal to give up old behavior and adopt a new response to a changed situation. The change in behavior is not the result of natural selection but is the beginning of freedom and free choice. Faced by a problem, an intelligent animal can find a solution and communicate it to the group by example and teaching. This solution is essentially behavioral, since the mode of conducting oneself is the only trait on which the will can have an impact and which occupies a place at the interface of the individual and his environment. When the animal faces challenge—climatic, predative, etc.—its response is primarily behavioral. If the animal is able to find an effective behavior response, there is no need for an organic response; thus, the species escapes natural selection.

Under favorable conditions, an observer may see the diffusion of a new behavior pattern in a monkey troop as a result of an experiment. Tsumori [230d] has given us an example. Japanese macacas like to eat tubers. One group living at Kyushu, close to the sea, dug up its tubers and rubbed them by hand to take off the dirt and gravel before eating them. Some monkeys noticed that the roots that fell on the beach were washed by the sea and had a salty taste, which the animals probably appreciated. They formed the habit of washing the tubers on the beach before eating them. Eventually, the whole troop adopted this new custom. Having extended their territory to the coast, the monkeys started eating the small crustaceans they found. Thus, the invention of a new technique was followed by an enlargement of the ecological niche.

Many other examples could be cited of the aptitude of primates for learning new behavior and transmitting it to their offspring. The totality of these learned patterns of behavior constitutes a *protoculture*; a species spread out over a large area may show regional protocultural differences. Such differences could correspond to the variations in selective pressure observed from one place to another.

This tendency finds its culmination in man, where it represents *culture*. It is because man acquired very early many learned behavioral patterns (weapons, clothes, shelter, fire, and, later on, agriculture and breeding, hygiene, and medicine) that he was able to occupy all the environments of the earth without splitting into different and specialized species. This progressive transition from inborn to acquired patterns of behavior can be perceived throughout the phylogenic ladder of the vertebrates. It is present in the birds, becomes more definite among the mammals, is still more evident in the primates, and reaches its full expression with man. If natural selection has retained it, it is because this transition, which constituted "ascension" toward conscience and liberty, offered a certain advantage.

Blood Groups of Primates, pages 11–22
© 1983 Alan R. Liss, Inc., 150 Fifth Avenue, New York, NY 10011

Taxonomy of Living Primates

Prosimii. 11
Anthropoidea. 14
   Platyrrhini 14
      Callitrichidae 14
      Cebidae 14
   Catarrhini 15
      Cercopithecoidea. 15
      Hominoidea. 20

We saw in the first chapter that, in the first half of the Cenozoic era, the order Primates split into a series of lineages still living today. Here, we follow the classification proposed in 1945 by Simpson [191a] and developed by Napier and Napier [152] and other authors, which is generally accepted today.

The most commonly accepted taxonomy of Primates (see Fig. 2-1) divides them into two suborders: Prosimii and Anthropoidea. This division, although far from being a natural one, has the important merit of separating "lower primates" into a single entity.

PROSIMII

As proposed by Simpson [191a], Prosimii have been divided into three infraorders, Lemuriformes, Lorisiformes, and Tarsiiformes, of which the first two have much more in common with each other than they do with the third. Both Lemuriformes and Lorisiformes are characterized by the so-called "dental comb," an arrangement of teeth that look and act like the teeth of a comb in grooming and feeding. While both Lemuriformes and Lorisiformes are popularly known as "lemurs," in the stricter sense this term is reserved for Madagascan lemurs that belong to three families of Lemuriformes: Lemuridae, Indriidae, and Daubentoniidae. The fourth family of the Lemuriformes, the Tupaiidae, are widely distributed throughout the Far East, as are also some representatives of Lorisidae, namely Nycticebus and Loris.

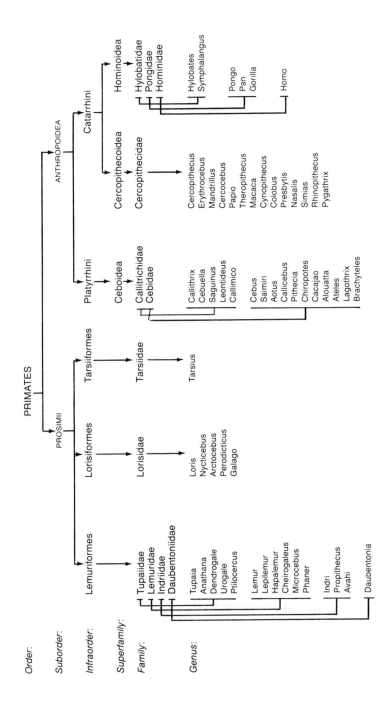

Fig. 2-1. Taxonomy of living primates.

The remaining genera of Lorisidae—Arctocebus, Perodicticus and Galago— are native to Africa.

Of the four families that constitute the infraorder Lemuriformes, the Lemuridae—and among those the animals that belong to the genus *Lemur*— are the best known and most commonly bred under captive conditions. All known (5–8) species of lemurs, *Lemur catta* (ring-tailed lemur), *L. variegatus* (ruffled lemur), *L. macaco* (black lemur), *L. mongoz* (mongoose lemur), and *L. rubriventer* (red-bellied lemur), are found in the major forest zones of Madagascar and the Comoro Islands. With a few exceptions, they are crepuscular animals with arboreal habitat, rarely being seen on the ground.

In the eastern parts of Madagascar, the lemurs share their arboreal habitat with a single representative of another family of Lemuriformes, Daubentoniidae, namely with the aye-aye (*Daubentonia madagascariensis*). This nocturnal animal with long tail and large eyes and ears has become very rare due to the extensive deforestation of Madagascar.

The third family of the Madagascar Lemuriformes, Indriidae, embraces three genera: *Avahi*, *Indri*, and *Propithecus*. These are small to medium-sized animals adapted to arboreal life, with a strictly vegetarian diet; they feed mainly on leaves and are consequently difficult to keep in captivity.

Tupaiidae, the tree shrews, constitute the fourth family of Lemuriformes and the only one with a geographical range outside Madagascar. Here belong *Anathana* (Madras tree shrew), bushy-tailed squirrel-like animals of India, south of the river Ganges; *Dendrogale* (smooth-tailed tree shrew), an animal the size of a large mouse, found in Vietnam, Thailand, and Cambodia; *Tupaia* (terrestrial tree shrew), a small squirrel-like primate living on the forest floor on the islands of the Malay Archipelago; *Urogale* (Philippine tree shrew), one of the largest of the tree shrews, which nests in holes in the ground or in cliffs of the Mindanao Island.

The Lorisiformes, another order of Prosimii, embrace a single family, Lorisidae, of which the best-known representatives are the African galagos or bushbabies. These small, largely nocturnal primates inhabit tropical rain forests between the Congo and Cross rivers; they thrive in captivity and are often kept as pets.

Two other, less common African representatives of the family Lorisidae are *Arctocebus* (angwantibos) and *Perodicticus* (pottos). The Asiatic branch of the same family is represented by two genera: *Loris* (slender loris) and *Nycticebus* (slow loris), both found in the tropical rain forests of Southeast Asia (India, Ceylon, Burma, etc.).

The infraorder Tarsiiformes embraces only one family, Tarsiidae, with a single representative, *Tarsius*, found on islands of the East Indies, e.g.,

Borneo and the Philippines. Because of some special features, such as reproductive characteristics and social habits, the classification of tarsiers has been continuously changed: they have been placed with Prosimii, in Anthropoidea, or in a group peculiar to themselves.

The tarsiers are primates about the size of a rat, with characteristically naked hands and feet and with terminal pads of their fingers and toes, greatly enlarged to form flat, soft disks with small flake-like nails. They are inhabitants of rain forests on many islands of the Malay Archipelago.

Since information dealing with the blood group serology of Prosimii is scant, if not nonexistent, a more detailed discussion of their taxonomy seems superfluous here.

ANTHROPOIDEA

The suborder Anthropoidea, or "higher Primates," is divided into two infraorders: Platyrrhini, the New World or South American monkeys, and Catarrhini, the Old World monkeys and apes, and man. The New and Old World groups share certain features that are more advanced than those of the Prosimii. Among others, all have short faces but large brains, their frontal and occipital lobes in particular being very large; the *corpus callosum* as well as associated areas are enlarged. Both groups have enlarged orbits separated from the temporal fossa by a nearly complete bony wall.

Although it is assumed that New and Old World monkeys have a very distant common origin, morphological differences between them are not significant. Platyrrhini have three premolars while Catarrhini have only two. The nostrils are far apart in South American monkeys compared to monkeys of the Old World, who have a narrow septum and nostrils looking forward and down. In Platyrrhini, the thumb is not opposable; objects are often gripped between digits 2 and 3.

Platyrrhini

Callitrichidae. Of the two families that constitute the infraorder Platyrrhini, the Callitrichidae, or marmosets, form a group of their own. These smallest of the South American monkeys have certain reproductive characteristics that set them apart from most of the Anthropoidea; e.g., they very often give birth to twins, which are carried on the father's back except during moments of nursing.

Cebidae. Another family of the New World monkeys, the Cebidae, is a diverse group that embraces small-size quadrupedal monkeys such as night monkeys (*Aotus*) and squirrel monkeys (*Saimiri*) and larger forms: capuchins

(*Cebus*), woolly monkeys (*Lagothrix*), spider monkeys (*Ateles*), and howler monkeys (*Alouatta*), all characterized by a prehensile tail that can support the body weight when the animal is hanging from a branch.

Catarrhini

The Catarrhini, or Old World higher primates, extend to Asia, Africa, and a very small part of Europe, and comprise two superfamilies: Cercopithecoidea (Old World monkeys) and Hominoidea (apes and man).

Cercopithecoidea. The Old World monkeys are quadrupedal animals with well-developed and often long tails and prominent ischial callosities. Their molar teeth are characteristically bilophodont, and the males have long canine teeth with sharp edges and a groove that extends to the root. Unlike the New World monkeys, the Cercopithecoidea, by and large, live in large social groups with well-marked rank order, particularly among males. The single family of the Old World monkeys, Cercopithecidae, comprises a number of genera with extensive laboratory use and long captivity records; for that reason, they are also extensively studied for their blood groups.

Genus* Papio *(the baboons). These are strongly built animals with long dog-like muzzles. They live in large troops numbering 50 or more that coalesce around a central group of dominant males. The alpha male mates with most females that are in estrus, so that the harem-like structure constitutes the nucleus of the troop. Baboons range from the southernmost tip of Africa to the subsaharan regions in Senegal in West Africa and to eastern Ethiopia, with small encroachment into southwestern Arabia.

The taxonomy of baboons proposed by Napier and Napier comprises two species groups. The larger, the *Papio cynocephalus* group, is divided into the following four species.

Papio anubis (olive baboon), which has the northerly range in subsaharan Africa, from Sierra Leone in the west to Central Ethiopia in the east (Fig. 2-2).

Papio cynocephalus (yellow baboon), which occupies the central geographical position, from northern Angola to northern Mozambique and further north to eastern Kenya and south to East Somalia.

Papio papio (Guinea baboon), the most westerly form, has its habitat in Sierra Leone, Guinea, and Senegal.

Papio ursinus (chacma baboon), the most southerly form, extends from the habitat of *P. cynocephalus* south to the Cape of Africa.

The smaller group, *Papio hamadryas*, comprises only one species: *Papio hamadryas* (sacred baboon). It is the easternmost form of the genus *Papio*, encountered in East Ethiopia, North Somalia, and Yemen and South Yemen on the Arabic Peninsula.

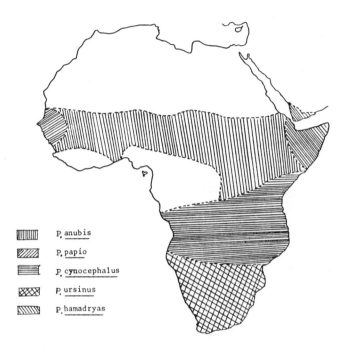

P. anubis	
P. papio	
P. cynocephalus	
P. ursinus	
P. hamadryas	

Fig. 2-2. Geographical range of baboons (modified from Napier and Napier [152]).

Genus Theropithecus (the geladas). Somewhat similar to baboons and geographically overlapping the range of hamadryas baboons, the geladas were formerly referred to as gelada baboons but now are considered a definitely separate genus that comprises one species and two subspecies. They live in large herds of up to 400 individuals on grassy mountain slopes at the edges of cliffs in central and north Ethiopia. Within large herds, smaller groups of animals are recognizable, which are reproductive units of several females and infants grouped around one breeding male.

Genus Macaca. The monkeys that belong to 13 species of genus *Macaca* cover an impressive geographical range from North Africa (Atlas mountains) and Gibraltar through southwestern (Afghanistan) and central (Tibet) Asia to China, Japan, and Formosa, and south to India, Ceylon, and throughout southeastern Asia, including Sumatra, Java, Borneo, Philippines, Celebes, and many offshore islands. Artificially introduced, the macaques can be found in Mauritius and Cayo Santiago (western India). With such a vast territorial range, they are adapted to a variety of climatic conditions and habitats from sea level to high altitudes in the mountains of Central Asia and North Africa, from tropical to montane forests, from swamps to grasslands and dry areas of scrub or rocky cliffs. These heavily built monkeys with strong limbs of almost equal length and with a rich coat, particularly in colder climates, usually live in large groups in which there is more than one adult male. Their social behavior varies from species to species and habitat to habitat.

Taxonomic classification of the macaques may be simplified by dividing them into two groups: short-tailed macaques and long-tailed macaques (C.P. Groves [62a]).

Short-tailed macaques have either no tails or little, stumpy ones projecting from between the ischial callosites. They include the following species.

1. *Macaca sylvanus* (Barbary apes or Gibraltar macaques) are the only macaques that live outside Asia and the only ones indigenous to North Africa. They seem, like the Japanese macaques, to be morphologically adapted to a snow-covered mountainous habitat.

2. *Macaca fuscata* (Japanese macaques) are native to Honshu, Shikoku, Kyushu, and Yakushima islands, have slightly longer tails than the preceding species, and can reach 18 kg (males) or 16 kg (females) in body weight, thus being the heaviest among the macaque species.

3. *Macaca thibetana* (Tibetan macaques) constitute a little-known species found in the mountains of Szechwan in southern China; they were previously confused with stump-tailed macaques, with which they can share the same habitat or inhabit neighboring areas in southern China.

4. *Macaca arctoides* (stump-tailed macaques), previously also known by the name *Macaca speciosa*, inhabit southern parts of China, Vietnam, Burma, and peninsular Thailand. Their most striking peculiarity is a curious genital specialization: the glans penis is very long and tapering with the urethral opening on its dorsal surface. According to Fooden (quoted by C.P. Groves [62a]), the vaginal anatomy of the females of this species is adapted in such a way to the shape of the penis that they form a "lock and key" effect, which would constitute an efficient mechanical isolation from other species.

5. *Macaca maurus* (or *nigra*) (Moor macaques or black or Celebes apes) inhabit Celebes and adjacent small islands. There are at least seven recognizable forms of Celebes macaques, each inhabiting a different geographical area. Some appear as intermediary forms, hardly distinguishable from monkeys that are sometimes classified as a separate genus, *Cynopithecus*.

The group of the long-tailed macaques comprises eight species, of which some are extensively used as laboratory animals and are frequently bred in captivity, so that data concerning their blood groups are relatively abundant.

6. *Macaca mulatta* (rhesus monkeys) is a species which, with its four subspecies, covers a wide geographical range from eastern Afghanistan through northern peninsular India, Nepal, Bhutan, Tibet, and provinces of southern China to Indochina and Thailand. Massive exportation of these monkeys caused significant depletion of wild populations in their natural habitats.

7. *Macaca fascicularis* (or *irus* or *cynomolgus*) (crab-eating macaques) are related to the rhesus monkeys and share with them common patterns of geographic variation; like the rhesus, they are widely used as laboratory animals. Their range extends from southern Indochina to Indonesia and the Philippines, where they are also known by the names Java monkeys, Phillipine macaques, or cynomolgus monkeys. At least 18 subspecies of crab-eating macaques have been recognized, with slightly different geographic ranges.

8. *Macaca nemestrina* (pig-tailed macaques) extend their range from Burma to Sumatra and Borneo; apart from rather small-sized separate races from Borneo and Bangka they have two widely recognized forms: the Malayan-Sumatran subspecies (*M. nemestrina nemestrina*) and the Burmese subspecies (*M. nemestrina leonina*).

9. *Macaca silenus* (lion-tailed macaques) is a rare species very closely related to pig-tailed macaques but restricted to relatively small areas in the hill forests of southwestern India.

10. *Macaca assamensis* (Assam macaques) superficially resemble the rhesus monkey and are native to Nepal, Assam, northern Burma, and the northernmost regions of Indochina.

11. *Macaca cyclopis* (Formosan rock monkeys), a rare species closely related to rhesus and crab-eating macaques, is found only in Taiwan.

12. *Macaca sinica* (toque macaques) is a species native to Ceylon, where each of its three subspecies occupies a different climatic zone, from low country to highland regions.

13. *Macaca radiata* (bonnet macaques) are found exclusively in southern India but are so closely related to the toque macaques of Ceylon that they might both be better classified as one species. Like the toque macaques, the bonnet macaques have anatomic peculiarities of the glans penis that may make interbreeding with other species difficult if not impossible.

Genus Cercopithecus *(vervet or green monkeys, guenon).* This is a widely spread group of monkeys found in the rain forest, woodlands, and savannah habitats of subsaharan Africa from Senegal in the west and Ethiopia in the east to the Cape in the south. The taxonomy of the genus *Cercopithecus* is still largely a matter of dispute, particularly as to the number of separate species and the number of "subspecies groups" within a single species.

According to a classification proposed by Hill (quoted by Napier and Napier [152]), there are nine "species groups" that include 21 species with 67 subspecies.

1. Aethiops group comprises grivets (*C. aethiops*), vervets (*C. pygery-thrus*), and green monkeys (*C. sabaeus*) found in very large areas of the African continent from Senegal and Ethiopia in the north to the open savannah of South Africa.

2. The Cephus group includes only one species, the moustached monkeys (*C. cephus*), which live an arboreal life in the tropical rain forests of Gabon and Congo.

3. The Diana group embraces three subspecies of Diana monkeys (*C. diana*), arboreal monkeys of West African rain forests, from Sierra Leone to Ghana.

4. The Lhoesti group includes l'Hoest's monkeys (*C. lhoesti*) and Preuss's monkeys (*C. preussi*), the two species of monkeys from Central (Congo) and East (Uganda) African lowlands and montane forests.

5. The Hamlyni group includes only one species of little-known owl-face or Hamlyn's monkeys (*C. hamlyni*) found in the eastern Congo basin in northwest Ruanda.

6. The Mitis group comprises blue (*C. mitis*) and Syke's (*C. albogularis*) monkeys, which are widely distributed throughout Africa, from Sudan to South Africa, and are characterized by an arboreal style of living within the great variety of forest habitats.

7. The Mona group includes Mona monkeys (*C. mona*), Campbell monkeys (*C. campbelli*), Wolf's monkeys (*C. wolfi*), crowned guenons (*C. pogonias*), and Dent's monkeys (*C. denti*), all arboreal inhabitants of the rain forests of West Africa from French Guinea to Cameroon and of Central Africa from Cameroon and Gabon to Uganda.

8. The Neglectus group comprises only one species, de Brazza monkeys (*C. neglectus*), with very wide distribution throughout East and Central Africa, mainly along banks of streams, on the trees as well as on the ground.

9. The Nictitans group is constituted of five species: spot-nosed guenons (*C. nictitans*), lesser spot-nosed guenons (*C. petaurista*), redtails (*C. ascanius*), red-eared guenons (*C. erythrotis*), and red-bellied guenons (*C. erythrogaster*). These are arboreal inhabitants of forest edges on the river banks of Cameroon, Congo, Kenya, Tanzania, Zambia, and Angola.

Previously included in the genus *Cercopithecus* but now often separated as the distinct genus *Miopithecus* are the smallest of the Old World monkeys, talapoins, also called mangrove monkeys. There is only one species of talapoins, *Cercopithecus talapoin*, which includes four subspecies, all living in west-central Africa along the Atlantic coast from southern Cameroon to Angola. Talapoins are easy to handle and to breed in captivity and thus are becoming increasingly popular as laboratory animals.

Genus **Erythrocebus.** This genus comprises only one species (with four subspecies) of patas monkeys (*Erythrocebus patas*), ground-living, relatively large animals from subsaharan Africa (Senegal, Sudan, and southward to Uganda). They are increasingly used as laboratory animals.

Genus **Presbytis (***langurs***).** Four "species groups" of langurs embracing 14 species are defined, of which the *Presbytis entellus* group is the best known, as the sacred monkeys of Hindus (*Hanuman* or *Entellus langur*). The others are the *Presbytis senex* group (purple-faced leaf monkey and John's langur), the *Presbytis aygula* group (Sunda Island leaf monkey, banded leaf monkey, white-fronted leaf monkey, and maroon leaf monkey), and the *Presbytis cristatus* group (silver leaf monkey, dusky leaf monkey, Phayre's leaf monkey, Francois' leaf monkey, Mentawai leaf monkey, capped langur, and golden langur). The langurs have a wide geographical range that includes India, Pakistan, Nepal, parts of Tibet, Sikkim, and Assam, and southward to Ceylon, Java, Borneo, Sumatra, Thailand, and many offshore islands of southeastern Asia. Their habitat is as diverse as their geographic range: from high mountain rocks (up to 3,500 m in the Himalayas), to dry arid zones of southern India and Ceylon and to rain forests and swamps of Indochina and Malaya.

Langurs belong to the Old World Monkey subfamily Colobinae, in which are classified also the African guerezas (*Colobus*). Several species of colobus monkeys are widely distributed across central, west, and east Africa.

Hominoidea. Members of the superfamily Hominoidea represent the highest grade of organization in primate phylogeny. In this category belong the family of the lesser apes of the far east (Hylobatidae), the family of the great apes of the Old World (Pongidae), and the family Hominidae, represented by a single worldwide polytypic species, *Homo sapiens*.

Lesser apes. The family of the lesser apes of the Far East comprehends two genera: *Hylobates* (gibbon) and *Symphalangus* (siamang).

Gibbons are confined to areas of primary forest in Southeast Asia. They are small (weights ranging from 4.0 to 8.0 kg for adult males and 4.0 to 6.8 kg for adult females), tailless apes with long, dense, shaggy fur varying from black to pale fawn and silver grey. Six species of gibbons are recognized: *Hylobates lar* (white-handed gibbons), found in Indochina, Thailand, the Malay Peninsula, and Sumatra; *Hylobates agilis* (dark-handed gibbon), which cohabits with the white-handed gibbon in Malaya and Sumatra; *Hylobates moloch* (silvery gibbon), the only species found in Java and Borneo; *Hylo-*

bates hoolock (Hoolock gibbon), found in Assam, Burma, and West Yunnan; *Hylobates concolor* (black gibbon), native to Vietnam and Laos; and *Hylobates klossii* (Kloss's gibbon), which is confined to the Mentawai islands off the west coast of Sumatra.

The genus *Symphalangus* comprises only one species: *Symphalangus syndactylus* (siamang). This is an ape more heavily built than gibbons (males reach 13.0 kg, females 11.6 kg), with a long and shaggy coat of uniformly black color, except for whitish hairs around the mouth and chin. Siamangs are found in Sumatra and the Malay Peninsula, in tropical rain forests as well as in montane forests at altitudes reaching 3,000 m.

Great apes. The family of great apes (Pongidae) comprises two genera of African apes: *Pan* (chimpanzee) and *Gorilla,* as well as one genus confined to southeastern Asia, *Pongo* (orangutan).

The commonly accepted taxonomy of chimpanzees recognizes the existence of two species: *Pan troglodytes* and *Pan paniscus,* the former having the following three (or even four) subspecies.

Pan troglodytes troglodytes, or Tschego (black-faced chimpanzee, also known in Europe as "common chimpanzee"), is found in Central Africa, between the Niger and Congo rivers.

Pan troglodytes verus (masked chimpanzee) (also called in Great Britain and the United States the "common chimpanzee") is found in forest regions of West Africa, from Sierra Leone to French Guinea.

Pan troglodytes schweinfurthii (eastern or long-haired chimpanzee) inhabits the regions of Central and East Africa from the Lualaba River eastward to Lake Victoria in the north and Lake Tanganyika in the south.

A fourth subspecies has been recognized by Hill (quoted by Groves), namely *Pan troglodytes koolookamba,* which shares the territorial range with *P. troglodytes troglodytes* but differs from the latter by such features as gorilla-like nose and sagittal crest.

Pan paniscus, or the pygmy chimpanzee, considered to be a separate species, is of lighter build than the other chimpanzees and has distinctly darker facial skin. Its geographical range is limited to enclaves formed by the Congo and Lualaba rivers.

The current classification of the genus *Gorilla* recognizes the existence of a single species divided into three subspecies:

Gorilla gorilla gorilla, or western lowland gorilla, is found in lowland rain forests in western parts of equatorial Africa, from the extreme southeast of Nigeria through Cameroon, Spanish Guinea, Gabon, Congo, Brazzaville, and the Central African Republic.

Gorilla gorilla beringei, or eastern highland gorilla, inhabits montane forests and open slopes up to altitudes of over 4,000 m in the Virunga

TABLE 2-1. Classification of primates proposed by Hoffstetter (after Szalay and Delsun [228])

Suborder: Infraorder:	Paromomyiformes = Plesiadapoidea:	Plesiadapiformes Paromomyidae (including Purgatorinae); Microsyopidae ?; Plesiadapidae Carpolestidae; Picrodontidae

Suborder:
Infraorders:
	Strepsirhini
	Lemuriformes
	  Adapoidea:
	  Lemuroidea:
	  (Daubentonioidea ?):
	
	Lorisiformes ?
	  Lorisoidea:
		
		Adapidae (including Notharinae)
		Lemuridae; Indriidae; Megaladapidae
		Daubentoniidae
		
		Lorisidae ?; Cheirogaleidae

Suborder:
Infraorders:
	Haplorhini
	Tarsiiformes
	  Tarsioidea:
	  (Omomyoidea ?):
	
	Simiformes
	Platyrrhini
	  Ceboidea:
	
	Catarrhini
	  Cercopithecoidea:
	  Hominoidea:
		
		Tarsiidae (including Microchoerinae)
		Omomyidae (including Anaptomorphinae)
		
		Cebidae; Xenothricidae; Callithrocidae
		
		Parapithecidae, Cercopithecidae
		Hylobatidae, Pongidae, Hominidae,
		Oreopithecidae

Volcanoes and the Mt. Kahuzi districts and the high mountain regions to the north and east of Lake Kivu.

Gorilla gorilla graueri, or eastern lowland gorilla, is found in the area east of the Upper Congo (Lualaba) river to the mountains west of Lake Edward and west of the northern tip of Lake Tanganyika.

Pongo (orangutan), the only great ape of Asia, is represented by a single species with two subspecies:

Pongo pygmaeus pygmaeus (Bornean orang) and the slightly larger, lighter-colored *Pongo pygmaeus abelii* (Sumatran orang) are both arboreal animals found at all levels of the tropical rain forests, where they feed among smaller branches but are only rarely seen on the ground.

On the basis of new biological data, R. Hoffstetter [66] has recently proposed a slightly different classification (Table 2-1), which divides the primates into three suborders: Paromomyiformes, which represent an early splitter whose existence was limited to the end of the Secondary and beginning of the Tertiary periods; Strepsirhini, which correspond to the majority of prosimians (Lemuriformes and Lorisiformes); and Haplorhini. The latter correspond to the simians (in the classical sense), to which the Tarsiformes were added also.

Blood Groups of Primates, pages 23–29
© 1983 Alan R. Liss, Inc., 150 Fifth Avenue, New York, NY 10011

3

Monkeys and Man

Monkeys in History 23
From Nonhuman to Human Primates 25
Utilization of Primates in Medicine 26

MONKEYS IN HISTORY

The tendency to humanize or indeed deify certain animals is as ancient as humanity. We are aware, for instance, of the role played by the bear in all the cultures of the Old World, as exemplified by the repeatedly encountered "Feast of the Bears," which has meaning, more or less magical, over a large area, starting from North Japan (the island of Hokkaido), crossing Siberia, the Alps, and Central Europe, and ending in the Pyrenees. But it was the monkeys that were the most fascinating for our ancestors. To such an extent do they resemble us in morphology, posture, mimicry, and behavior that one could see in them parahuman individuals, situated beside or indeed above us. The ancient Egyptians worshipped baboons. Monkeys occupied a place in Greek and Roman mythology, and even today the langurs of India are holy; they are identified with the god Hanuman, who, with the support of his attendant monkeys, helped Rama (one of the avatars of Vishna) to conquer the king of the demons.

In China and Japan, monkeys are often the heroes of legends or fables in which they have the leading roles. In Madagascar, the nocturnal prosimians were held by the natives to be the reincarnated dead, and were respected and sometimes dreaded. The attitude of the West, which became Christian, was quite different from that of Asia: here, most often, our inferior cousins were considered ludicrous or grotesque, symbolizing ugliness and deceit.

During the European Middle Ages and the Renaissance, monkeys were caricatured. In folklore, art, and tradition they represented all the vices of man—falsehood, lust, vanity—and were mocked. They were dressed in human garments at fairs; at royal courts they were compelled to stand upright and handle objects, to the great amusement of the people and the princes.

As for the great anthropoid apes, discovered fairly early by explorers, they were taken for a long time to be a special species of human beings like fauns, satyrs, and sphinxes, and as such captured the popular imagination. During the third century B.C., the Carthaginian Hanno brought back from a voyage to the West Coast of Africa skins of great apes, no doubt chimpanzees although he called them gorillas, which he had taken for hairy human beings living in a savage state. He wrote: "The company arrived at an island full of savage inhabitants, of whom most were females, thickly covered with fur, which our interpreters called gorillas. Three females were captured . . . they did not accept our companionship. Having killed them, we skinned them and carried their skins with us to Carthage."

In the first century of our era, Galen methodically dissected, for the first time, the African macaque and showed its close anatomical resemblance to man, which was to be more precisely described some hundreds of years later by the Belgian anatomist Vesalius. During the last part of the seventeenth century, the English surgeon Tyson undertook the detailed anatomical study of the chimpanzee, which he called orangutan or *Homo sylvestris*.

In addition, in spite of the progress in knowledge during this epoch the boundary between the human and the nonhuman remained fluid; witness Tyson's admirable sketch, in his famous work, showing a young chimpanzee upright, a cane in his hand, looking straight before him. The eighteenth century teems with engravings in which one sees monkeys, alone or in a family group, elegantly dressed, using the most sophisticated musical instruments of the times. In fact, it was only toward 1758, when the 10th edition of *Systema Naturae* was published, that Linnaeus systematized the classification of the primates and assigned to man a place exactly beside the anthropoid apes, forming together one similar group. Even though he was deeply religious and never invoked the idea of transformation, the scheme of Linnaeus created a scandal. It again placed man—a being created in the image of God and the angel fallen through original sin—by the side of the great apes, who, intelligent as they were, nonetheless remained, for the people of this era, animals bereft of morals and carriers of all the vices. But an even greater shock was to come one hundred years later, in 1859, with the publication of Darwin's famous work, *On the Origin of Species by Means of Natural Selection, or the Preservation of Favoured Races in the Struggle for Life*, which was completed in 1871 by *The Descent of Man and Selection in Relation to Sex*. Henceforth, not only is man a primate, a neighbor of gorillas and chimpanzees, but now the naturalists recognize in him an ancestry shared with the monkeys. The periodical discovery of fossil remains came to describe precisely the forms of transition in support of those theories in which the branch of the anthropoid apes separated from the human branch in the course of time. In addition, these discoveries revealed the evolutionary stages passed through by our lineage from the "initial bifurcation" to "*sapiens.*" From this time on, the common origin of man and the monkeys could no longer be doubted.

FROM NONHUMAN TO HUMAN PRIMATES

The morphological resemblances, bony and visceral, between apes and man, which were impressive to the anatomists of the past centuries, are met again, in a striking fashion and perhaps even accentuated, at the cellular and molecular level. The karyotypes of the different species of primates are closely related, as shown by the authoritative studies of Dutrillaux [36]. When they are compared to the human karyotype, Robertsonian translocation among the Lemuridae and fusions among the Cercopithecidae are found. When one compares the Pongidae with man, one finds pericentral inversions and a terminal rearrangement. The latter results in the formation of one mediocentric chromosome from the two acrocentric chromosomes. In a similar way, a karyotype of Pongidae that contained 48 chromosomes became a human karyotype with only 46 chromosomes.

It was believed for a long time that chromosomal rearrangements occurred at random. In fact, Dutrillaux et al. [37] have recently shown that, at least in primates, certain chromosomal zones underwent changes more often than others, which remained surprisingly stable. It was within those "sensitive" zones that all modifications are found when karyotypes of two closely related species are compared. Significantly, those same "sensitive" zones are the most frequent sites of chromosomal accidents that occur when cell cultures are exposed to irradiation.

At no time, however, does the euchromatin show the major changes involved in the type of deficiencies or duplications that would imply a loss or gain of chromatin material. *Banding* methods permit the reconstitution of different chromosomal sequences, which fundamentally remain the same in all species. The same relationship is found again at the level of the gene products—the enzymes, hemoglobins, and immunoglobulins, the HLA histocompatibility groups, and of course the red blood cell groups, with which this work is concerned.

Whatever system is studied, one notes in many species related, sometimes even identical, factors; and when differences exist, they do not appear to have any effect on the function of the molecule. However, the modifications observed permit us to establish the phylogenetic distances or intervals between species and to trim the evolutionary tree.

UTILIZATION OF PRIMATES IN MEDICINE

At the time of the Renaissance, and during the two centuries that followed, observations of physiological phenomena in animals opened the doors to scientific medicine. Thanks to this comparative method, Harvey discovered the circulation of blood, which he published in 1628. But it was necessary to wait for Claude Bernard (1813–1878) for a clear formulation of the methods of "experimental medicine." During this era and until the eve of World War II, scientists devoted themselves particularly to the study of the "macrofunctions": respiration, circulation, locomotion, endocrine and exocrine secretion, etc. It was found that the mammals were all structured on the same model in which structure and function corresponded to a body of genetic information, fundamentally identical. In addition, for as long as one remained at the level of macrofunctions that covered gross processes, the mechanisms that one studied changed little from one species to another. What was observed in rodents or the carnivores was also seen in primates and man. During that era, one resorted exclusively to the use of the dog, cat, rabbit, rat, mouse, and guinea pig as laboratory animals.

With the progress of science, as investigations were gradually leaving the level of gross macrofunctions to reach that of molecular microfunction, it was necessary to call upon those species that were phylogenetically closer to man. The separation of the great mammalian stocks that we know today must go back to the very beginning of the Tertiary geological era, which was 70 million years ago. The separation of anthropoid apes and man occurred no more than 15–20 million years ago. Also, since they were separated so much later from the same original branch, we can conceive that *Homo sapiens* has maintained a much larger portion of DNA sequences in common with primate animals than with groups further removed phylogenetically, such as the carnivora, the rodents, and the Lagomorpha.

The nonhuman primates, especially the anthropoid apes, can contract the human diseases, whether degenerative, viral, bacterial, parasitic, tumorous, or even behavioral. Indeed, the course of these diseases and the resulting illnesses evolve very much as they do in man. Thus, they constitute experimental models that are remarkably suitable for studying human pathology.

While establishing itself on the biochemical, immunological, and molecular levels, experimentation with primate animals was becoming more sophisticated and refined. Its application in areas such as pharmacology, immunology, and endocrinology called for the use of tissue or cell culture techniques rather than bloody surgical intervention involving the whole body of the animal.

In this way some new areas of interest to several essential branches of contemporary medicine could be explored by using nonhuman primates. This occurred first in immunology, because many primates have antigenic systems closely related to, or even identical with, those of man. The resemblance is such that one can remove the blood of a chimpanzee or even a baboon and replace it with human blood. Thus, one obtains an animal whose circulating blood is a human tissue. This permits the investigation *in vivo* of the action of certain viruses on human cells or the sensitivity of the same cells to certain drugs. Such investigation would be impossible to achieve in man for obvious ethical reasons. The use of animals zoologically further removed from us than the higher primates is hardly possible for this type of research because the injected human blood would be destroyed almost immediately by the heterospecific antibodies. We shall cite a recent example. As is well known, there is a serious health problem connected with the high-frequency occurrence of the sickle cell trait in certain African populations; yet, few medications are available that are able to ameliorate the symptoms of anemia and especially the articular symptoms associated with this malady. Oswaldo Castro, at the Howard University College of Medicine, has successfully investigated *in vivo* the sickling phenomenon, as well as the action of certain antisickling agents, by infusing chimpanzees and baboons with isotope-labeled human sickle and normal red blood cells and by comparing their survival rates.

This tolerance of the higher primates for the red blood cells of man has made possible in certain cases the establishment of a cross-circulation between man and monkey. In some particularly severe forms of viral hepatitis, the establishment of such a circulation between the patient, in hepatic coma, and the baboon, previously exchange-transfused with human blood, permits the monkey to take charge of the hepatic metabolism of the two organisms for the time necessary for the liver of the patient to recover its function.

The close immunological kinship linking the higher primates to man explains why many human infections (tuberculosis, cholera, streptococcemias, trepanematosis, venereal diseases) can be transmitted to certain nonhuman primates. Often they show the same type of evolution, thus providing irreplaceable experimental models for the study of pathology, the use of antibiotics, and the establishment of therapeutic regimens.

A similar situation exists for many of the parasitic diseases, especially malaria, which today afflicts the greatest number of human beings. The *Plasmodia*, which parasitize man, are endowed with a strict specificity. They are not found in other mammals, aside from some nonhuman primates. Thanks to the use of monkeys, we have been able to study the nature of immune reactions stimulated by *Plasmodia* and the causes of sensitivity to the infection. Other experiments have dealt with the immunology of schistosomiasis in monkeys.

It is above all in virology that the utilization of nonhuman primates has advanced our knowledge. Many viruses are rigidly specific and live only in man and in closely related species of apes, the chimpanzee and the gibbon, and, exceptionally, in other species of primates.

The poliomyelitis virus was first cultivated in monkey kidney tissue. More recently, the chimpanzee has been used to study the biology of viral hepatitis of type B and non-A/non-B and to produce the first experimental vaccines. Chimpanzees and gibbons are used by the manufacturers of the antiviral vaccines for quality control of their products.

The viral etiology of kuru (an illness found in New Guinea and in the neighboring islands in people who customarily ate the brain of deceased members of their family in order to acquire their spirit and experience) was established by inoculating chimpanzees with the brains of individuals who had died of this disease. After several years, the animals showed the typical symptoms of kuru. This proved the existence of a new form of virus, called slow-virus, capable of causing death after a very long and latent period of the development of the disease.

The study of certain viruses carried spontaneously by several species of monkeys (above all the New World monkeys) revealed their oncogenic potency. Two viruses of this group have been the object of particularly intensive investigation: *herpesvirus saimiri*, and *herpesvirus ateles*. Both viruses were found to be carcinogenic, but there were significant interspecies and even individual differences in the susceptibility observed. Moreover, it was found that these viruses may be pathogenic to animals beyond the group of primates: they caused tumors in rabbits. They had not been found, however, in man, and they differ in many respects from the Epstein-Barr virus (EBV), which occurs relatively frequently in our species but whose exact pathogenic role remains to be elucidated. Nevertheless, there are some basic similarities between EBV and the two herpesviruses of primates—in particular, the ability of both kinds of viruses to induce cellular transformation. EBV is regularly observed in certain types of cancer with strict geographical distribution, such as the nasopharyngeal cancer of Southeastern Asia and Burkitt's lymphoma of black Africa. This is a concordance that merits investigation in depth and that makes one believe that the oncogenic viruses, now well identified in nonhuman primates, could in the future provide a study model applicable to man.

The same may apply to studies focusing on the leukemias. Leukemia-like disease has been induced in monkeys by injections of materials obtained from leukemic patients.

Similarly, the close physiological kinship among all higher primates, including man, makes the nonhuman primates the model of choice in the investigation of the etiology and prevention of metabolic disorders. Primate animals share with man high requirements of many nutrients and vitamins. Osteomalacia, rickets, macrocytic anemia, and several other vitamin deficiency diseases are common in monkeys and can be easily induced experimentally. Cardiovascular changes such as atherosclerosis and arteriosclerosis can also be produced in primate animals by dietary manipulation. Nutrition liver lesions, such as fatty liver, were also found in monkeys to be associated with a deficiency of lipoproteins. As primates also require a dietary intake of high-quality proteins, they may serve as an experimental animal model for the investigation of protein malnutrition. Important contributions to the knowledge of the toxicology of chronic alcoholism and of the pathology of alcoholic liver cirrhosis were provided by the use of baboons as experimental animals in the study of alcohol-abuse–related diseases.

The similarity of the reproductive organs of monkeys to those of man make the monkeys an irreplaceable model for the study of the biology of reproduction and fertility, as well as for the development of methods of contraception. They also constitute ideal models for research on neurological processes, such as sleep and the origin and nature of circadian rhythms—problems of great importance for the organization of nightworker schedules and the timetables of airline pilots [161]. They are successfully used in research on the processes of aging and in the identification of the biological basis of the acquisition of learned behavior, symbolical communication, etc.

Finally, the ethological and psychological relationships between the nonhuman primates and man make the former superior to other animal models. They are situated, so to speak, at a stage intermediate between man and the more ancient groups, such as carnivores and rodents. Certain of those have been widely used for a long time and are thus well known in many respects, but since they detached themselves from the common mammalian branch very long ago, they are of little use for the study of some functions, especially those highly developed in the human.

In contrast, the nonhuman primates separated themselves from us at a more recent date and have conserved in their genetic patrimony certain archaic traits that are witnesses of the stages through which our distant ancestors, who disappeared long ago, had to pass. From this point of view, the nonhuman primates constitute for the anthropologist and the evolutionist a true "time machine."

Blood Groups of Primates, pages 31–38
© 1983 Alan R. Liss, Inc., 150 Fifth Avenue, New York, NY 10011

History of the Discovery of Blood Groups in Monkeys

Discovery of the First Human Blood Groups   31
Immunological Polymorphism of Living Organisms   32
Discovery of Blood Groups in Domestic Animals.   35
Discovery of Blood Groups in Nonhuman Primates   36
Problems of Nomenclature   38

DISCOVERY OF THE FIRST HUMAN BLOOD GROUPS

It was Landsteiner, an Austrian scientist (born in Vienna on June 14, 1868, and died in New York on June 25, 1943), who, in 1900–1901 at the age of 32, discovered the first blood groups in man [85]. Based on patterns of cross-agglutination observed when red blood cells of certain subjects were mixed with the sera of others, Landsteiner distinguished two antigenic factors that he called A and B. According to whether these factors were present simultaneously or separately, or were altogether absent from the red cells, four types of blood, A, B, AB, and O, were identified. The antibodies present in the serum, on the other hand, were found to be directed against antigen(s) absent from an individual's own red cells (Landsteiner's rule). The reciprocal relationships between the red cell antigens and the antibodies in the serum in the four A-B-O blood groups are shown in Table 4-1.

In 1908, Epstein and Ottenberg [42a] put forward the hypothesis that these factors were hereditary. This was soon confirmed by Von Dungern and Hirszfeld [32a,33]. The model of genetic control of these factors consisting of a series of three allelic genes—two codominant, A and B, and a recessive, O—was established by Bernstein [7a,b] in 1924 and 1925 and subsequently confirmed by other authors.

This fundamental discovery, which earned Landsteiner a Nobel Prize 30 years later (1930), had many consequences. For the first time, characteristics had been discovered that were directly or almost directly controlled by clearly

TABLE 4-1. Constitution of the four basic blood groups of the A-B-O system

Blood group:	A	B	AB	O
Antigens present on the red blood cells	A	B	A and B	Neither A nor B
Antibodies present in the serum	Anti-B	Anti-A	None	Anti-A and Anti-B

Only the first three groups, A, B, and O, were discovered by Landsteiner; the fourth, AB, was found one year later by Decastello and Sturli [246].

identified genes. Human genetics and physical anthropology were to be profoundly modified by these findings.

In fact, until then, researchers had been caught between two contradictory approaches. Morphological traits (upon which all traditional anthropology was based) proved to be an unreliable source of information as they do not depend on a single gene but, most probably, on highly complex models, which therefore excludes them from any strict analysis. Moreover, they are largely influenced by environmental conditions. The other approach was based on the acceptance of simple mutations that seemed to correspond to a single factor (hemophilia, agenesis of a finger, etc.). But these are virtually always pathological and, indeed, sometimes even monstrous conditions, which fortunately are extremely rare and because of their low frequency can hardly be used for statistical analysis.

At this time, human genetics seemed to be limited to a "horror show." Landsteiner's discoveries changed everything. The blood groups, with their frequent occurrence and their monomeric character that escaped all environmental influences, proved to be free from all the drawbacks of the previously studied hereditary traits. Thanks to blood groups, the mathematical models of population genetics became applicable also to the human race.

IMMUNOLOGICAL POLYMORPHISM OF LIVING ORGANISMS

The discovery and study of blood groups revealed the existence of genetic polymorphism in man, i.e., the presence of hereditarily controlled immunological differences among subjects belonging to the same race or the same population. At the time, this was an entirely new and even "shocking" concept, which deserves a more detailed examination.

From earliest antiquity, men used to classify living creatures in groups of identical individuals or at least those with a certain number of basically identical characteristics. The basis of this classification is the species, which assembles individuals who all carry the same specific features. It, in turn, can be divided into groups of individuals demonstrating the same racial features. Linnaeus was the first naturalist who, in the middle of the eighteenth century, added some precision to this method. During that period, knowledge was associated with identification and classification. It was on this method of linnaean typology that zoologists, botanists, and anthropologists of the nineteenth century and the first half of the twentieth century based their science. This typological vision of the world conformed to the thinking of Darwin, who considered that all variation occurring within a homogeneous group was a rare and precarious phenomenon, as natural selection must make a rapid choice between the "normal" and the "mutant," in order to preserve only that which is best adapted to the pressures of the environment. However, it was the spreading of blood transfusion that showed the full extent and ubiquity of human polymorphism. The idea of practicing blood transfusion is probably very old, and chroniclers record many attempts made in Europe, particularly since the Renaissance. In some cases, transfusion gave spectacular results, but more often than not it caused the death of the recipient. In fact, everything depended on the compatibility of the donor's blood with the serum of the recipient, which, because of the ignorance of blood groups at the time, was entirely a matter of chance. The discoveries of Landsteiner made it possible for transfusions to be performed in relative safety. It seemed sufficient to select a donor whose blood lacked the antigen that could be destroyed by the antibodies present in the recipient's serum. Nevertheless, it was only in 1914, when it became known how to conserve blood rendered incoagulable by the addition of small quantities of sodium citrate, that transfusion became common practice. (The method of conserving blood was discovered independently and almost at the same time by Hustin in Belgium, Agote in Buenos-Aires, and Lewissohn in New York.) From 1917 on, the allied armies engaged in World War I widely practiced blood transfusion, thus helping to save many lives. At the end of the war, blood transfusion became a routine procedure; it assured the important progress in surgery that marked the following decades. During the last years of the war (1917-1918) Ludwik and Hanna Hirszfeld, serologists of Polish origin, found themselves with the Serbian armies on the Salonika front. The great allied powers such as France and Great Britain sent into this sector reserves of troops drawn from all over their territories and colonial empires. Working on subjects of various ethnic origins, the Hirszfelds [65a] discovered that all types of blood groups were found in all races and that no population existed that was exclusively group A or group B or group O. Only the frequencies of the blood types were found to vary from one point of the globe to another. These variations were virtually always of a progressive and continuous character.

This was completely opposed to the typological way of thinking, which postulated the existence of categories made up of identical subjects. In reality, all populations are composed of individuals who demonstrate quite a wide genetical diversity. The conclusion that assumed the existence of variations within a single population and replaced the typological belief based on the notion of the uniformity of the population probably constituted the most important conceptual revolution to have occurred since Charles Darwin.

Later, many other blood groups were discovered: first the M-N and P systems by Landsteiner and Levine in 1927 [87,88] and the Rhesus system by Landsteiner and Wiener in 1940 [98]. Others followed: Lutheran, Kell, Lewis, Duffy, Kidd, Diego, Y^t, Auberger, Dombrock, etc. All are polymorphic, i.e., they arise from one or several series of different mutations. It is thanks to this polymorphism that they are detectable by immunological methods: a locus that is strictly monomorphic obviously cannot give rise to the phenomenon of alloimmunization. Alloimmunization is the introduction of an antigen into a subject who does not possess it. This can take place in two cases: first, during blood transfusion, when blood containing a certain factor is injected into a recipient who himself lacks this factor; second, during pregnancy, if the fetus carries an antigen that the mother lacks. In each case, alloimmunization is shown by the appearance of "irregular antibodies" active against the antigens that elicited their formation. (Regular antibodies appear "regularly" in the serum [for example, anti-A, anti-B] according to the blood group; the irregular antibodies [anti-D, anti-M, anti-Kell] are exceptional and generally appear after stimulation.) They are, therefore, found either in polytransfused subjects who have demonstrated reactions despite the observance of the traditional rules of compatibility or in multiparous women who have given birth to children with hemolytic diseases of the newborn.

The large majority of "new" blood group systems thus described displayed a high degree of polymorphism. As early as 1939, W.C. Boyd published the first atlas of the distribution of blood groups in man [12]. It included all the systems then known: A-B-O, M-N, and P. In 1954, A.E. Mourant drew up maps of the distribution of the factors that were discovered later (Rh, etc.) [145]. His revised and supplemented work on worldwide distribution of blood and serum groups appeared in 1976 [146].

In the 1960s, the introduction of the technique of electrophoresis made possible the study of the genetic polymorphism of a number of proteins, particularly those that form the enzymatic groups. Soon the available data became sufficient to draw the main outlines of a geographical hematology that covered the principal human populations of the five parts of the world [6].

From 1966 on, Lewontin and Hubby [110] applied the techniques of electrophoresis to other groups of living organisms, in particular to the arthropods. They demonstrated the constant presence and the extent of polymorphism encountered in all populations, animal as well as vegetable. Polymorphism appears today as a fundamental attribute of life.

DISCOVERY OF BLOOD GROUPS IN DOMESTIC ANIMALS

Attempts to detect by immunological methods individual differences among animals, particularly among domestic animals that were readily accessible to investigation, started quite early and, in fact, even preceded Landsteiner's discovery of the A-B-O blood groups of man. In 1900, just before he published his observations, Ehrlich and Morgenroth [41a] drew attention to the existence of individual differences in the blood of goats, demonstrated by immunological methods. At that time, Hirszfeld had suggested that comparable differences might be found in man, but the research he envisaged had not yet been started when Landsteiner published the results of his studies.

In 1910, von Dungern and Hirszfeld described the first blood groups in dogs (for reference, see Irwin [69]). The same year, Todd and White [230c] reported the presence in cattle of polymorphic blood systems that they had observed during vaccination against cattle plague.

In 1921, Hirszfeld and Przesmycki [65a] discovered horse blood groups; in 1926, Szymanowski and Wachler [228a] found those of pigs; and in 1930 Todd [230a,b] found those of chickens. Before and after World War II, numerous studies were carried out on the blood groups of domestic animals. This knowledge was used by breeders to trace pedigrees [32,68]. The development of methods of artificial insemination further increased the interest in immunogenetic identification.

Today, the study of blood groups constitutes part of routine breeding procedure. It enables us to trace the lineage of a family, in an objective and precise manner, in order to improve the quality of the breeding stock we wish to develop. It is also possible to increase the output of certain domestic animals (milk/meat production) at the cost of an overall impoverishment of the genetic patrimony and an increase in monomorphism. But a group that has lost part of its alleles becomes more susceptible to exterior pressures and often requires far more care and closer attention than the polymorphic group of the wild populations.

What is gained on one side by developing an economically profitable characteristic is lost by diminishing the possibilities of reaction by the group to external hazards. A pure race, obtained by the breeder, is always fragile and would have little chance of survival away from the artificial conditions of the man-made environment.

The advantage of genetic polymorphism, found in all natural populations, is precisely that it increases the capacity to meet the multitude of requirements of the enviroment. The polymorphic group possesses a wide range of responses and can therefore adapt itself to situations that the monomorphic group cannot endure. That is why natural selection has always favored the maintenance or increment of genetic polymorphism, which is certainly linked to the constant changes, in space as well as in time, of the environmental conditions in which we live.

It is probably the same mechanism that explains the exceptional vigor of the descendants of parents who differed greatly in their genetic patrimony. The highly heterozygous offspring collect a number of qualities inherited from each of the two parents. This phenomenon, called *luxuriance of hybrids* or *heterosis,* has often been put to use in agriculture.

DISCOVERY OF BLOOD GROUPS IN NONHUMAN PRIMATES

Bearing in mind the close phylogenic affinities between man and nonhuman primates, it is understandable that the discovery of blood groups in man prompted early attempts by immunologists to find in monkeys the blood group systems that were known in man. However, this undertaking was far more arduous than the investigations of domestic animals; primate animals remain less numerous and less accessible to investigators. Moreover, they reproduce slowly (one offspring per gestation) and mating is difficult in captivity.

In the earliest attempts to detect blood groups on the red cells of nonhuman primates, human antisera were used from which the anti-monkey heteroagglutinins were eliminated, first by elution, and later by absorption with appropriate erythrocytes. In that way, Landsteiner and Miller [93,96] were able to demonstrate, as early as 1925, that chimpanzees belonged to either O or A groups (the factors[1] being present both in the saliva and on the red blood cells). Subsequently, A and B factors were found in secretions of both the Old and New World monkeys, while the presence of those factors on the

---

[1]In the field of red cell groups, immunologists construct genetic models on the basis of phenotypes, defined by serological methods, since it is not yet possible to gain access directly to the genes as is the case, for example, with simple globin chains. Thus, for the purpose of simplification we have used throughout the text the term "factor" to define both the gene and its product on the red cell.

red cells was found to be an attribute of the higher primates only, namely of anthropoid apes and man. On the other hand, not all the primate species were found to display the presence of all the A-B-O factors simultaneously. Some species have three factors, others only two, while still others have only one, thus demonstrating genetic monomorphism within the A-B-O group system [243,245]. As for the M and N antigens, their presence was also confirmed on the red cells of anthropoid apes, who also were shown to possess blood factors related to the Rh system.

All these factors present both in man and in nonhuman primtes were for a long time called "human-type blood factors" [260,261]. This terminology demonstrates a certain anthropocentrism from which immunologists could no more escape than could anatomists and physiologists. In reality, the blood types present in nonhuman primates and man existed millions of years before our species appeared on earth.

A good many other studies, covering different fields, have been carried out during the past 25 years by Alexander Wiener's team and by his pupil Jan Moor-Jankowski, who was instrumental in founding the Laboratory for Experimental Medicine and Surgery in Primates (LEMSIP) at New York University.

Earlier, we mentioned that many blood groups had been discovered thanks to the appearance of irregular antibodies in the sera of polytransfused and multiparous subjects. There are, all over the world, thousands of polytransfused individuals and hundreds of thousands of multiparous women who have all been the subject of immunological screening. Not so for monkeys, who, in natural conditions, receive neither transfusions nor medical check-ups for pregnant females. The majority of cases of alloimmunization in the course of incompatible pregnancies (which must surely occur) escape the notice of the observer. Wiener and Moor-Jankowski's method was to create the conditions of alloimmunization artificially by repeatedly injecting one animal's blood into that of another of the same species.

Thus, in a certain number of experiments they were able to produce immune antibodies that recognized factors present on the donor animal's red cells but absent from the recipient's blood cells. This study, long and meticulous, allowed them to define a number of the so-called "simian-type blood groups," which did not correspond to any known antigens of man. In the next phase of the experiments, the monkeys were immunized with human red cells, and the cross-reactions obtained with the resulting antisera were used to study the relationships between the simian-type blood groups and those of the so-called human type. This approach proved to be quite useful. It demonstrated that many factors considered purely simian (which they really are if only their zoological distribution is considered) actually belong to genetic systems common to man and certain nonhuman primates, some fractions of which have evolved differently in different species.

We will see in the course of the following chapters numerous examples of this *diversifying evolution* that affects an immunological system.

For the same reason, instead of opposing the human-type blood groups (encountered in man as well as in various nonhuman primate species) to the simian-type blood groups (which, in turn, are the endowment of some species of the primate animals), it appears more in conformity with the current data on immunogenetics to discuss the problems of blood groups of nonhuman primates in terms of *blood group systems.*

PROBLEMS OF NOMENCLATURE

Finally, another difficulty should be considered, i.e., that posed by nomenclature. The majority of factors were named as they were discovered, even before it was known to which system they belonged. The labels they received depended therefore on the chronology of of their discovery and were in fact due to chance. When Wiener and Moor-Jankowski described the first antigens resulting from the experimental alloimmunization of the chimpanzee, they named them alphabetically, according to the order of their discovery: A^c, B^c, C^c, D^c, E^c, F^c, etc. (the exponent c indicating that it was an antigen present in the chimpanzee). Later, it was shown that A^c, B^c, D^c were related to the M-N system, whereas C^c, E^c, and F^c were related to the Rhesus system. However, the first two had nothing to do with the A and B factors of the human A-B-O system, while C^c, D^c, and E^c must not be connected to factors C, D, and E of the human Rhesus system. Despite the presence of the exponent (c), which was meant to avoid ambiguity, this virtual homonymy does not make things very clear to the nonspecialist.

That is why, as soon as knowledge of the subject advances further, it will be useful for all workers in this field to convene and to settle on a universal nomenclature that would end the existing confusion. Such a revision would be indispensible to facilitate future work. It should take into account the relationship of each factor to a particular genetic system, and should follow certain rules accepted by the majority of immunologists. It is timely to recall that the acceptance of such general rules in relation to the very complex HLA system helped to avoid confusion similar to that which for a long time surrounded and plagued the nomenclature of the Rh system. This would allow clarification and unification of the blood groups of nonhuman primates, as was done for the human blood groups. Such an undertaking should be easy because the research teams that devote their activity to the red cell groups of nonhuman primates are not numerous. Throughout the world they can be counted on the fingers of one hand.

Blood Groups of Primates, pages 39–51
© 1983 Alan R. Liss, Inc., 150 Fifth Avenue, New York, NY 10011

The A-B-O System

The Nature of the Antigens of the A-B-O System 39
The Genetic Model of the A-B-O System. 39
The Subgroups of A: A$_1$ and A$_2$ 45
Variations of A-B-O Polymorphism in Nonhuman Primates. 51

THE NATURE OF THE ANTIGENS OF THE A-B-O SYSTEM

We will not return to the history of Landsteiner's discovery of the A-B-O system at the beginning of the century, as this has already been thoroughly dealt with in the preceding chapter.

Since then, this system has been the object of thousands of studies. We have described how, in 1924, Bernstein developed the genetic theory that supported the A-B-O system. After World War II, the biochemists Morgan, Watkins, and Kabat [238] clarified the nature of the antigenic receptors and showed that the system was fundamentally formed of three factors: H, the true basic substance, which results from the attachment of an L-fucose to a hydrocarbonate chain, the precursor, and to which can be attached either an N-acetyl-galactosamine, which expresses A specficity, or a D-galactose, which corresponds to B specificity. Recently, Oriol and collaborators [154a–154d] reported new data to explain the nature of the relationship between group substances present on the red cells and those found in secretions. We will discuss this subject later in the chapter. For a general review of the subject see Ropars and Cartron [163].

THE GENETIC MODEL OF THE A-B-O SYSTEM

Although the A, B, and H substances are present in many living species, even in very ancient ones, it is only in man and probably other primates that they are known to depend on a genetic system comprising at least two series of independent alleles [212,213]. The first is the *H/h* pair. The *H* gene ensures the synthesis of fucosyl transferase, which adds an L-fucose to the precursor. Its recessive mutation, *h,* leads to the appearance of the Bombay type in the homozygote.

The second is a polyallelic series, comprising, as Bernstein proposed, at least three mutations, *O, B,* and *A* (*A* being divided, in the majority of great apes and in man, into two fairly frequent mutations: A_1 and A_2). The *O* gene is silent, and if it appears alone in the genotype, the red blood cells carry only the H substance. The *A* and *B* genes are codominant, each controlling the synthesis of a corresponding transferase. Carriers of the *A* gene produce a specific transferase that adds N-acetyl-D-galactosamine to the precursor; gene *B* adds D-galactose instead, as noted above.

Bernstein's model applies not only to man; it fits almost all simian primates: New World monkeys (Platyrrhini) and Old World monkeys (Catarrhini), who very regularly show A, B, or H factors in their secretions, depending on their genotype; and anthropoid apes (gibbons and orangutans from Asia and chimpanzees from Africa), in whom these factors are also simultaneously present on the red blood cells.

In almost all cases, antibodies are present in the serum for which the corresponding agglutinogens are absent from the red cells or from the secretions (Landsteiner's rule).

The presence of this system in all species of primates demonstrates that it must have already existed in the common ancestor, which lived in the Oligocene era, some 60 million, or perhaps more, years ago.

If Bernstein's model now seems established in the phylum that produced the simians (since all still living species display it), the synthesis of similar but more archaic factors, now called B-like and A-like, can be observed in the older groups.

Many prosimians and New World monkeys carry a B-like antigen on their red blood cells, as do the Old World monkeys, as Landsteiner noticed in 1911. However, its synthesis is controlled by genes independent of those which constitute Bernstein's system. Although very similar to the salivary and erythrocytic factors B, the B-like antigen differs somewhat from them serologically, and its presence in an individual does not preclude the appearance of anti-B antibody in the serum (see Fig. 5-1).

In light of observations made in nonhuman primates, we may conclude that the development of the A-B-O blood group system as it appears in man probably proceeded in several stages, which may be reconstructed as follows.

1. The very early (i.e., at the bacterial stage) appearance of genes capable of coding for the transferases that ensure the presence of A, B, and H substances (or very closely related substances) on the cellular membranes or in organic fluids [70]; and

2. much later, the establishment of an allelic relationshp (according to Bernstein's model) among the *A,B,* and *O* and then among the A_1, A_2, *B,* and

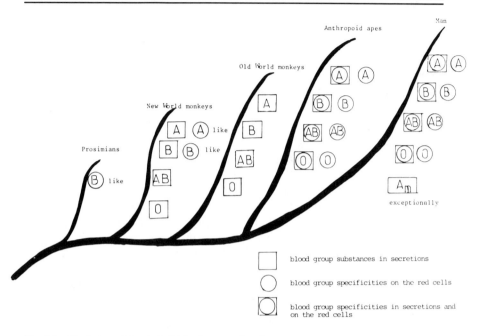

Fig. 5-1. Evolution of the A-B-O blood groups in primates.

O genes, corresponding to the simian stage (since this genetic system is displayed by all simian species, including the most archaic—for instance the New World monkey species).

3. Since, as mentioned earlier, the factors controlled by the *A, B,* and *O* genes of the Old and New World monkeys as well as those of prosimians are found only in secretions but never on the red cells, the next step is the passage of blood group substances onto the erythrocytes, a modification that appeared with the anthropoid apes and persists in man. It is assumed that this passage is linked with the action of a *Y* gene responsible for the presence of an A specificity on the red cells and of another, comparable gene whose intervention is necessary for the appearance of the B substance on the red cells [239]. There are rare human subjects, presumed to be homozygous for the recessive allele *y (y/y)*, who have the factor A only in their saliva but not on the red cells. Individuals of that type were described for the first time by Wiener and Gordon [249], who assigned to that type the symbol A_m (subscript "m" for monkey).

A_m blood imitates somewhat, in those rare humans of genotype *y/y*, an ancestral state that preceded the appearance of the anthropoid apes (and therefore could have existed in a unique state until the middle of the Miocene period, 15–20 million years ago). A comparable mutation, B_m, was recorded for group B.

All anthropoid apes, on the other hand, possess A-B-H factors both in their saliva and on their red blood cells. However, in the gorilla the quantity of antigens present on the red blood cells is considerably lower than in the other anthropoid species. The bulk of the blood group substance is present in the gorilla secretions. From the point of view of the immunogenetics of the A-B-O system, gorillas have remained at an earlier prehominoid stage.

As far as the secretor status is concerned, apes are genetically monomorphic, since they are all of the secretor type. They can possess only the *Se* gene. Nevertheless, some rare orangutans are nonsecretors. These were the first observed to possess, although admittedly in a very low frequency, the *se* mutation, which is far more widespread in the human species; about a third of all Europeans have a nonsecretor phenotype (i.e., an *se/se* homozygous genotype).

For the *H/h* allelic pair, the situation seems more complex. In fact, unlike group *O* humans, whose red blood cells are rich in H substance, the red cells of *O* group chimpanzees, for example, are not agglutinated by anti-H lectin *(Ulex europaeus)*. The same applies to the A_2 group, rich in H in man but not agglutinated by anti-H in the chimpanzee. In gibbons, anti-H lectin strongly agglutinates group B cells but gives only a weak or negative reaction with the group A or AB red cells (group O has not yet been encountered in this species). In contrast to gibbon red cells, the red cells of all orangutans, irrespective of blood group, fail to react with anti-H reagents.

Observations made in anthropoid apes are difficult to reconcile with the commonly accepted hypothesis that the H substance is a precursor of A- and B-group–specific substances, i.e., that the H determinant must first be attached to the molecule of precursor substance before the molecule will accept the A and B determinants at their appropriate sites. This hypothesis also falls short of giving a satisfactory explanation for the wide range of reactions that anti-H reagents produce with red cells from different individuals of group A_1 or group B, or for differences in the H reactions of the red cells of the same A-B-O blood groups from different races [56,268]. The alternative hypothesis, proposed by Wiener and his collaborators [283], assumed that the genes *A* and *B*, and gene *H*, all operate in parallel, on a common substrate, by attaching their respective determinant groups at their appropriate locations on the same single macromolecule. The reciprocal relationship between H and A-B would be, according to this interpretation, on the phenotypic level, and would depend on steric interference between the respective terminal groups due to their proximity to one another. If one can assume that the

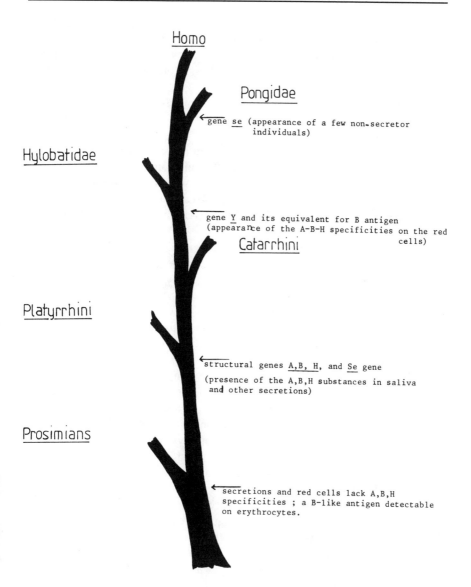

Homo

Pongidae

← gene <u>se</u> (appearance of a few non-secretor
individuals)

Hylobatidae

← gene <u>Y</u> and its equivalent for B antigen
(appearance of the A-B-H specificities on the red
cells)

Catarrhini

Platyrrhini

← structural genes <u>A,B, H</u>, and <u>Se</u> gene
(presence of the A,B,H substances in saliva
and other secretions)

Prosimians

← secretions and red cells lack A,B,H
specificities ; a B-like antigen detectable
on erythrocytes.

Fig. 5-2. Successive appearance of the genes of the A-B-O blood group system in the course
of immunologic evolution of primates. From Ruffié and Socha [187a].

precursor substance is different in the various human races and in various nonhuman primate species, the relative positions of the A-B-H determinant groups could very well also be different, thus accounting for various patterns of H reactivity.

It seems likely that biochemistry, and particularly enzymology, will play a decisive role in elucidating the true nature of the relationship between the *H* and *A-B* genes [17].

The study of the transferases of the human red cells of types A_m and B_m, as well as comparative investigations of the chimpanzee and human red cells of group O that have recently been initiated, will perhaps throw some light on the mechanisms of different genes involved in the biosynthesis of group substances, both on the red cells and in secretions, that existed at different evolutionary stages.

These stages can be represented by the diagram shown in Fig 5-2. The tree trunk stands for the phylogenic line, and the branches growing on either side correspond to the currently living groups; they have been curtailed for simplification.

If one adheres to the Morgan and Watkins view of the relationship of the H to the A and B factors, one can conceive the following stages in the development of the A-B-O system until it achieved the form currently found in primates.

1. The appearance of a polyallelic series composed of three fundamental structural genes, *A, B,* and *O* (the last being recessive and silent); the first two direct the synthesis of the basic factors, A and B, from the H substance (which, in turn, is controlled by an allele on an independent locus).

2. The establishment of several regulatory systems, at least two of which are known in detail: The first is *Se/se,* of which the gene *Se,* dominant and virtually fixed (monomorphic) in all prehuman species, assured the formation of the A and B antigens in their glycoprotein (hydrosoluble) form, present in the saliva and other secretions; the recessive mutation, *se,* appears only exceptionally in anthropoid apes but is quite frequent in man. The second is *Y/y* for the A antigen and, probably, another, equivalent pair for the B. These genes assure the formation of the antigens A and B.

According to Oriol and collaborators, who recently came up with a new concept of the genetics of the A-B-O, *H-h,* secretor, and Lewis systems [154a–154d], the A-B-H antigens in secretions and those on the red cells have separate origins and different structures. There are two types of precursor chains: *type 1,* with terminal β-D-Gal-(1,3)-β-D-Glc-Nac-R; and *type 2,* with terminal β-D-Gal-(1,4)-β-D-Glc-Nac-R. Type 2 would correspond to specificity H found in vascular endothelial cells, while type 1 is characteristic of the H antigen present in external secretions, in plasma, and on the lymphocytes. The red blood cells carry types 1 and 2 antigens, but only that of type 2 is of endogenous origin; type 1 antigen is assumed to be secondarily affixed on erythrocyte membranes from the H substance in plasma (acquisition of the Lewis phenotype). Thus, type 1 antigens are synthesized exclusively in the cells of mesodermal origin, while type 2 antigens are the product of the cells of epithelial origin. Contrary to the previous belief that the *Se* gene was a regulatory gene that controlled the expression of the structural *H* gene in external secretions, Oriol et al. consider *Se* and *H* genes to be two distinct but closely linked structural genes, the consequence of a duplication of an ancestral *H* gene. The two genes could code for distinct α-2-L-fucosyltransferases, one reacting preferentially with type 1 chains, and the other reacting preferentially with type 2 chains.

THE SUBGROUPS OF A: A_1 and A_2

The division of human group A into two types, A_1 and A_2, was proposed in 1911 by von Dungern and Hirszfeld [33]. It results from the existence of two autonomous mutations that are also encountered in anthropoid apes, but here the immunological relationship is not exactly the same as that observed on the human level [29].

Following the discovery of subgroups of A, the difference between A_1 and A_2 was thought to be purely quantitative. But in 1927 Landsteiner and Witt [99] demonstrated, by repeated absorptions and warm elutions, that the difference between the two subgroups was also qualitative. The factor initially called A was in fact composed of two antigens: A, present in A_1 as well as A_2 individuals; and A_1, found only in A_1 individuals. The general assumption was that the A_2 represented a "less mutated" form of A, as shown by the fact that by and large the A_2 red cells reacted more strongly with anti-H and more weakly with anti-A than the red cells of A_1 type. The same inverse relationship can be observed between the quantity of H substance and that of B substance. The more the cells contain of A and B, the less the H, and vice versa.

Since then, this difference in the nature of the two A antigens has been confirmed by a large number of authors (for review, see [27]). For Wiener and Socha [281] the factors A_1 and A_2 were differentiated by the length of the subjacent oligosaccharide chains carrying the antigenic determinant, the A_1 chain being longer than the A_2 chain. It is now assumed that both types, A_1 and A_2, carry the same terminal composed of three or four sugars. What differentiates one subgroup from another is the relative proportion of straight

TABLE 5-1. ABO blood groups of apes and monkeys

Species	O	A	B	AB	Total number tested	Remarks
ANTHROPOID APES [a]						
Chimpanzees						
Pan troglodytes	70	651	0	0	721	Subgroups A_1 and A_2 detectable
Pan paniscus	0	14	0	0	14	All type A_1
Gibbons *(Hylobates)*						
(various species)	0	27	60	56	143	Subgroups A_1, A_2, A_1B, and A_2B detectable
Siamangs *(Symphalangus)*	0	0	2	0	2	
Orangutans *(Pongo pygmaeus)*	0	48	14	22	84	Subgroups A_1, $A_{1,2}$, A_2, A_1B, $A_{1,2}B$, and A_2B detectable
Gorillas						
Gorilla g. gorilla	0	0	42	0	42	Based on saliva inhibition and
Gorilla g. beringei	0	0	4	0	4	serum tests, red cells very weakly reacting with anti-B
OLD WORLD MONKEYS[b]						
Baboons						
Papio anubis						
Ethiopia [13]	0	8	146	78	232	Gene *O* presumed to be present
Kenya [259]	0	53	56	65	174	Gene *O* presumed to be present
Papio hamadryas						
Ethiopia [209]	0	15	107	50	172	Gene *O* persumed to be present
Hybrids *(hamadryas/anubis)*						
Ethiopia [209]	0	5	81	43	129	
Papio cynocephalus						
Kenya [259]	0	18	20	22	60	Gene *O* presumed to be present
Papio papio						
Senegal [259]	2	27	93	66	188	
Papio ursinus						
Cape Province [259]	0	4	59	26	89	Gene *O* presumed to be present
Natal [23]	0	11	10	10	31	Gene *O* presumed to be present
Transvaal [28]	0	22	4	22	48	
Papio (species unknown)	1	45	70	68	184	
Geladas *(Theropithecus gelada)*	20	0	0	0	20	Anti-A and anti-B not always detectable in serum
Macaques						
Macaca arctoides	0	0	41	0	41	

TABLE 5-1 (Continued)

Species	\	Blood group	\	\	Total number tested	Remarks
	O	A	B	AB		
Macaca fascicularis						
Malaysia [229]	4	111	64	66	245	
Indonesia [229]	2	91	26	46	165	
Philippines [229]	2	3	100	25	130	
Captive, Japan [229]	0	31	56	88	175	Gene *O* presumed to be present
Captive, USA	5	67	94	44	210	
Macaca fuscata	0	0	18	0	18	Based on serum tests only, saliva not tested; red cells strongly reacting with anti-B sera
Macaca mulatta	2	2	194	2	200	
Macaca nemestrina	66	14	7	2	89	
Macaca radiata						
India [144]	0	11	14	0	25	
Captive, USA	0	21	14	17	52	
Macaca sylvanus	0	32	0	0	32	
Drills *(Mandrillus)*	0	4	0	0	4	
Langurs *(Presbytis entellus)* [144]	0	0	16	2	18	
Celebes black apes *(Cynopithecus)*	1	23	2	0	26	Expected agglutinins not always detectable in the serum
Patas monkeys *(Erythrocebus)*	0	26	0	0	26	
Vervet monkeys *(Cercopithecus)*						
Ethopia [71]	0	119	0	1	120	
Bulge River [120]	0	16	5	1	22	
Natal [120]	0	19	8	4	31	
Natal [27]	0	18	8	8	34	
Transvaal [27]	0	21	2	2	25	
NEW WORLD MONKEYS[b]						
Spider monkey *(Ateles)*, various species	1	10	4	0	15	
Squirrel monkeys *(Saimiri)*	2	6	0	3	11	
Capuchins						
(Cebus albifrons)	1	0	3	0	4	
(Cebus apella)	0	5	0	0	5	
Howler monkeys *(Alouatta)*	0	0	52	0	52	
Marmosets *(Callithrix)*, various species	0	45	0	0	45	

[a]A-B-O groups established by hemagglutination and serum tests.
[b]A-B-O groups established by saliva inhibition and serum tests.

and ramified chains. According to Hakomori et al. [63], subgroup A_1 is characterized by a larger proportion of branched chains than subgroup A_2. Nevertheless, there must also be a quantitative difference: several measurements have confirmed that red cells of the A_1 type possess on their membranes far more antigenic sites than do erythrocytes of type A_2 [278]. However, the qualitative nature of the A_1/A_2 difference was again invoked in some later studies [122], thus reviving the concept of Landsteiner and Witt [99].

The problem becomes even more complex when one considers the properties of the A antigen in various human populations and in various species of nonhuman primates:

1. In blacks an A_{int} (intermediate) mutation exists, described for the first time by Landsteiner and Levine [91a], and precisely defined by A.S. Wiener [240a]. It can also be encountered, although exceptionally, in whites. The A_{int} red cells react more strongly with anti-A and anti-H than the A_2 cells do (which demonstrates that the A and H antigens are not necessarily complementary to each other on the red cell membrane [8]).

2. The subgroups A_1 and A_2 exist in at least three species of anthropoid apes: chimpanzees, orangutans, and gibbons. The studies of A.S. Wiener et al. [285] demonstrated that there is no strict inverse relationship between the quantity of H substance and that of A substance present on the red blood cells, in these species at least, if this quantity is measured by immunological methods. Indeed, if one consideres the reactivity of group A blood cells of various origins with anti-A reagent, one obtains the following gradations: A_1 human > A_1 *P. paniscus* > A_1 *P. troglodytes* > A_2 *P. troglodytes* > A_2 human. All these factors support the notion of a qualitative difference between factors A_1 and A_2, as was demonstrated in man by the biochemists Kabat, Yamashita et al. [293], and Hakomori et al. [63].

According to Hakomori et al. the glycolipids that take part in the formation of group A belong to four different biochemical types, A^a, A^b, A^c, and A^d, that can be separated by chromatography. On the other hand, the glycolipids that compose factor H also belong to four types: H_1, H_2, H_3, and H_4.

There is possibly a connection between the four types of the H series and those of the A series. Indeed, it is the transformation of H_1 that gives A^a, that of H_2 that gives A^b, that of H_3 that gives A^c, and that of H_4 that gives A^d. In each case, a specific transferase seems to intervene. It is the degree of transformation of these types that differentiates A_1 and A_2. In fact, the components of A seem far less abundant in group A_2 than in group A_1. This is particularly true for A^c and A^d, the latter fraction possibly even being absent from A_2, whereas it is consistently found in A_1. In contrast, it is the

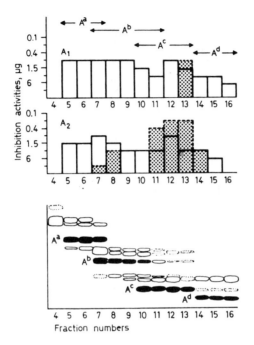

Fig. 5-3. Distribution of A and H activities over A and H variant glycolipids from group A_1 and A_2 erythrocytes. The lower panel shows the thin-layer chromatographic pattern of glycolipids of eluates from DEAE-cellulose columns. Solid spots = A^a, A^b, A^c, and A^d glycolipids. Open and dotted spots = nonactive glycolipids. The upper panel shows the blood group activities expressed as the minimum quantity (in μg) of glycolipid that can inhibit three hemagglutination doses of anti-A antibody and *Ulex* anti-H lectin. Open columns = A activity; dotted columns = H activity. From Hakomori et al. [63], with permission of S. Karger, Basel.

reverse phenomenon that occurs for the components of H; the H_3 and H_4 fractions are present only in individuals of group A_2 and not in A_1 individuals.

Thus, the difference between A_1 and A_2 would be both *quantitative,* there being a less complete transformation in A_2 of the four H factors into their four homologues, and *qualitative,* this difference in the level of transformation involving in particular the last two terms, H_3/A^c and H_4/A^d, which are mostly transformed in A_1 and far less so in A_2. Thus, the enzyme or, more probably, the enzymatic group that determines the appearance of A_1 is qualitatively different from that which determines the appearance of A_2.

In Hakomori's scheme, the transferases present in A_1 and A_2 are assumed to be capable of converting the H_1 and H_2 glycolipids into A^a and A^b (although the conversion in A_2 is weaker). In contrast, the conversion of H_3

and H_4 into A^c and A^d can be accomplished only by the enzymes carried by A_1 individuals but not by A_2 individuals. This is why A_2 contains more H_3 and H_4 determinants than A_1 but far fewer A^c and A^d determinants.

The proportions of these different fractions present on the red cell vary during ontogeny. In particular, the H_1/A^a and H_2/A^b variants are more frequent in the newborn than in the adult, who is more likely to display the types of H_3/A^c and H_4/A^D. Figures 5-3 and 5-4, reproduced by courtesy of Hakomori and co-workers, summarize this mechanism.

It appears, therefore, that there is a real immunogenesis that follows ontogenic development in the same way as with the other morphological or physiological traits. This immunogeneisis would reveal itself by a progressive

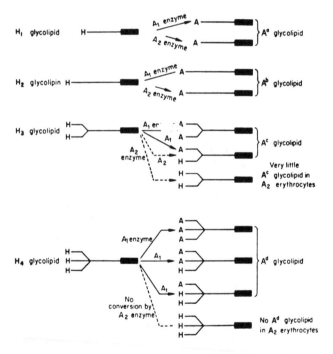

Fig. 5-4. Schematic representation of H and A glycolipid status in group A_1 and A_2 erythrocytes. The straight-chain H_1 and H_2 glycolipids can be converted to A^a and A^b glycolipids through both A_1 and A_2 enzymes, although the conversion through A_2 enzyme is less than through A_1 enzyme. The branched-chain H_3 and H_4 glycolipids can be converted to A^c and A^d glycolipids through A_1 enzyme, but not readily through A_2 enzyme. Consequently, group A_2 erythrocytes contain more H_3 and H_4 glycolipids and/or A^c or A^d glycolipids with incomplete A determinants. The structures of H_4 and A^d glycolipids are hypothetical. From Hakomori et al. [63], with permission of S. Karger, Basel.

elongation and dendrification (or ramification) of the molecule that characterizes each of the four stages defined above.

Finally, *other variations could appear during cancerous processes,* and the
malignant cells would show a considerable increase in their quantity of
structure 4 (H_4A^d) compared with normal cells (occasionally with a slighter
increase of the H_3 structure).

The same type of research must now be carried out in nonhuman primates,
especially in the chimpanzee, the orangutan, and the gibbon, in which two
erythrocyte groups, A_1 and A_2, exist that are not exactly comparable with
those of man.

Knowledge of the distribution of the four fractions of H and A in different
types may allow us to learn the way in which groups A_1 and A_2 developed
during the course of evolution, and whether the biogenetic principle of Fritz
Muller, "ontogeny recapitulates phylogeny," can also be applied in the field
of immunology.

VARIATIONS OF A-B-O POLYMORPHISM IN NONHUMAN PRIMATES

The A-B-O system is not always as polymorphic in nonhuman primates as
in man. All human populations (other than the pure-bred Amerindians of
South America) possess three fundamental genes, *A, B,* and *O,* even when
one of these genes is present in very low frequency, as in the case of the *B*
factor in the Basque population and, to a lesser degree, in all of southwest
Europe. This situation occurs in numerous species of primates but not in all,
as can be seen in Table 5-1. Nevertheless, some of these results must be
considered provisional because of the small numbers of animals tested in
certain species.

Blood Groups of Primates, pages 53–74
© 1983 Alan R. Liss, Inc., 150 Fifth Avenue, New York, NY 10011

The M-N System

The Genetic and Phylogenic Complexity of the M-N System 53
The M-N Series 54
    Review of Discoveries. 54
    The M-N Blood Group System in Man and Anthropoid Apes 56
    The Phylogenic Evolution of the Chromosome Segment Affecting the
    M-N System. 57
    The Genetic Model of the M-N System According to Wiener et al. . . . 59
    Mutations Specific to Each Species 63
    The Current Genetic Model of the M-N System. 65
The Henshaw (He) and Hunter (Hu) Antigens 66
Antigens of the Miltenberger Series 67
Antigens of the S/s/U Series 68
Antigens of the $V^c/A^c/B^c/D^c$ Series 69
Antigen W^c. 70
Phylogeny of the M-N Blood Group System 71

THE GENETIC AND PHYLOGENIC COMPLEXITY OF THE M-N SYSTEM

The A-B-O system, discussed in the preceding chapter, appears genetically simple and phylogenically stable. The structural genes that control the appearance of the group substances correspond to a simple series of alleles. Moreover, these factors have undergone no fundamental change from the arrival of the first simians to the appearance of man. The antigenic structures and the genetic model that controls them seem to have been only marginally modified, if at all, in the course of millions of years. They have probably remained unchanged since the beginning of the Tertiary era. The relatively minor changes that have occurred involve the reciprocal content of the H factor in relation to A and B factors and the appearance of the antigens on the red cells of the anthropoid apes. None of those minor variations implies any profound changes in genetic information. This relative stability in time seems to distinguish the A-B-O system from other blood group systems and from the M-N system in particular.

The M-N system is composed of the following factors.

1. Factors common to man and the primates closest to man. These are fractions of the M and N, Henshaw, and Mi^a antigens that, on the evolutionary level, correspond to what are called *paleosequences*.

2. Factors that seem to be specific to each phylum: S, s, U for the human lineage; A^c, B^c, and D^c for the anthropoid apes. They must correspond to *neosequences* that have each evolved in a separate direction.

This distinction between the two categories of antigens seems to be valid for the Rhesus system also, as we shall see in the following chapter.

Actually, it is difficult to draw a precise dividing line between these two categories of factors; some blood groups occupy pivot positions and create a "bridge" between human factors and simian factors. They are found at a kind of immunological crossroads in evolution. For the M-N system, this is the case of the N^V factor, which is present both in man and in many anthropoid primates, where it is closely associated with the V^c factor. The latter was first discovered in chimpanzees as a result of experimental alloimmunization.

For the Rhesus system, this is the case of the human Rh_0 antigen, which possesses a certain number of antigenic factors also encountered in the R^c antigen of the chimpanzee, gorilla, etc. A good many other antigenic crossroads must exist, but they may not appear until antisera become available that are sufficiently potent to react with primate red cells.

We will now consider these different categories of blood groups and will then try to define their phylogenic relationships.

THE M-N SERIES
Review of Discoveries

We have already indicated that the M-N system was discovered in 1927 by Landsteiner and Levine, who were attempting to demonstrate the existence of erythrocyte factors other than A and B. For this purpose, they injected human red blood cells into rabbits, hoping to provoke the formation of antibodies that, unlike the anti-A and anti-B, do not spontaneously occur in man. By means of these immune sera they expected to discover antigens hitherto unknown.

Reagents produced in that way enabled detection of three new antigens: M, N, and P, which appeared to be genetically independent of the A-B-O system. M and N were found to belong to a single system and to correspond to two codominant alleles, while P was presumed to be a dominant trait belonging, together with its recessive silent allele, to another, completely autonomous system [87–90].

It should be recalled that during a comparable experiment, carried out some years later, in which Landsteiner and Wiener injected into rabbits, not human, but rhesus monkey *(Macaca mulatta)* red blood cells, an immune antibody was produced that permitted the definition of another polymorphism in man. This discovery led to the description of the Rhesus system, the full importance of which is now known.

The M-N system, as first conceived by Landsteiner and Levine, was composed of three genotypes, corresponding to three phenotypes, as shown in Table 6-1.

In 1969, Jack et al. [70a] demonstrated that factor M contained a fraction M_1 that reacted specifically with serum anti-M_1. The factor M_1 is more frequent in blacks (24% positive among U.S. blacks and 50% positive among Bantus) than in whites (only 3–4% positive). The factor M_1 appears to have several degrees.

It was discovered quite early that M and N antigens were present in several primate species. In fact, the serological reactions observed in primate animals were not exactly comparable to those observed in man; they only showed some similar sequences. In 1937, Landsteiner and Wiener [97] discovered that some anti-M sera, prepared by the immunization of rabbits with human red blood cells, agglutinated the red blood cells of rhesus monkeys. Later, they extended those studies to other species, particularly to chimpanzees, which remain the most thoroughly studied primate species [240]. This comparative research demonstrated that at least six similar, although not identical, specificities could be recognized by testing red cells of different species with batteries of anti-M immune sera prepared by injecting group M human red blood cells into rabbits, as shown in Table 9-7.

Moor-Jankowski, Wiener, and Gordon [131] went even further and studied anthropoid apes that also presented a certain M-N polymorphism. In addition to a common M specificity, they managed to find a specificity particular to the gibbon, two specificities belonging to the chimpanzee—$M_1{}^{ch}$ and $M_2{}^{ch}$—

TABLE 6-1. The M-N system as conceived at the time of its discovery

Reactions with rabbit antiserum			
Anti-M	Anti-N	Phenotypes	Genotypes
+	−	M	*MM*
+	+	MN	*MN*
−	+	N	*NN*

and the M specificities peculiar to man. The way in which these factors, varying with the zoological group, combine suggests (when the phylogenic family tree is considered) a step-like evolution [171a]. We shall return to this concept.

Unfortunately, things were not so simple for the anti-N antisera, which always contain powerful heterospecific agglutinins that render the reading of the results far more uncertain. This is why in practice lectin extracts of *Vicia graminea* and *V. unijuga* are used instead of rabbit immune antisera.

The M-N Blood Group System in Man and Anthropoid Apes

At present, only anthropoid apes have provided sufficient information for useful comparison with the M-N system in man.

The available data demonstrate that differences exist that concern first the nature of the antigens. We have just seen that, besides the specificities common to all species, there exist specificities that are unique to each stock, for example the M factor in man, in the chimpanzee, and in the gibbon. But these differences also affect the mode of action of the genes. In man, the system consists of two codominant genes, *M* and *N,* each of which is capable of controlling the synthesis of the corresponding factor. This model is encountered only in man and the gibbon. In other African anthropoids (chimpanzee and gorilla), an allele is found that is capable of simultaneously controlling the synthesis of M and N [240,254,263,274].

However, it is possible that this is a difference of degree and not of substance. When we look closer, it seems that the immunological segregation between M and N is not absolutely precise even in man.

Actually, if the *N* gene leads to the exclusive appearance of the N factor on the red blood cell, the human *M* gene also ensures not only the presence of the M specificity but also traces of N.

Thus, when a rabbit is injected with human M red blood cells, it reacts not only by producing anti-M antibodies, but also weak anti-N antibodies. Likewise, the titer of an anti-N antibody subjected to repeated absorptions by group M red blood cells diminishes progressively. These effects were demonstrated by Landsteiner and Levine in 1928 [89,90]. They can also be observed when the anti-N of *Vicia graminea* [107] and *Vicia unijuga* [123] lectins are used as reagents.

This perhaps also explains the differences in the occurrence of natural anti-M and anti-N antibodies in our species. While some natural anti-M agglutinins are found in group N individuals (although this is rare), the presence of anti-N antibodies in M individuals is quite unusual. Moreover, these anti-N antibodies are easily absorbed with group M human red cells [65]. Only

M S−s−U+ or M S−s−U− types appear to be totally lacking in N factor (see below).

Let us now examine in more detail the differences observed between the M-N system in man and in anthropoid apes.

All chimpanzees carry an M factor. About 40% are also N-positive, and 60% are N-negative [254]. This factor N is detected only by *Vicia graminea* lectin. Chimpanzees, therefore, belong to two serological phenotypes, M and MN; the N type does not exist in these animals (Table 6-2).

All gorillas carry the factor N (also identified by *Vicia graminea)*, but 83% have the factor M, and 17% are M-negative (also called m). They display, therefore, two phenotypes, N and MN, group M being absent in this species [270].

Finally, 59% of orangutans carry the M factor. Since they never carry factor N, only two M-N phenotypes can be distinguished in this species: M-positive and M-negative. This is the only group of anthropoids that shows the equivalent of group O for the M-N system [266].

The Phylogenic Evolution of the Chromosome Segment Affecting the M-N system

The serological reactions observed when the M and N factors in man and in nonhuman primates are studied show that, despite some differences affecting certain antigen patterns, the same fundamental genetic system is involved. Its establishment must to a large extent have preceded hominization. Where,

TABLE 6-2. The M and N blood group factors in anthropoid apes

Species	M-N types encountered	Frequency	Number of animals studied
Chimpanzees			
Pan troglodytes	M	60.5	500
	MN	39.5	
Pan paniscus	M	100.0	14
Gorillas			
Gorilla gorilla	N	16.7	18
	MN	83.3	
Gibbons			
Hylobates lar	M	21.7	60
	MN	25.0	
	N	53.3	
Orangutans			
Pongo pygmaeus	M	58.6	29
	m	41.4	

then, do the differences in the mode of distribution of the two factors within the species studied originate?

Let us consider Table 6-2. If the genetic model generally accepted as explaining the inheritance of the M-N system in man (two codominant genes, one directing the appearance of N, the other that of M) is retained, it must be assumed that, at least in the chimpanzee and gorilla, the same gene can simultaneously ensure the synthesis of M and N (in any case, far more than the gene *M* does in man). In short, in these two primates, an "MN" mutation exists whose action brings about an end-product situated between the human M and N factors.

Such a mutation must be similar to M^c, described in 1953 by Dunsford et al. [34], which simultaneously ensures the appearance of M and N antigenic patterns on the red blood cell. In short, M^c reproduces the situation encountered in chimpanzees and gorillas. (Nevertheless, M^c red blood cells, considered intermediary between M and N by these authors, are not agglutinated by *Vicia graminea* lectin.)

Other very rare variants, M^r and M^z, composed of M and N patterns have been described. M^r and M^z react like M^c with the anti-M and the anti-N antisera, but different sera are involved in each case, which demonstrates that each mutation possesses its own specificity. It is possible that the genes responsible for the appearance of M^r and M^z correspond to *reserve mutations* and reestablish in man a situation comparable to that existing in African anthropoid apes (gorilla and chimpanzee).

On the biochemical level, factor N represents the precursor of M [214,215,231]. We may conclude that the *M* and *N* alleles of the two African anthropoids, chimpanzee and gorilla, have not gone as far in the transformation of N into M as has the *M* gene in man. In other words, we may be dealing here with less-mutant forms of *M,* which have preserved a much greater capacity to synthesize N than the *M* mutation in man. On the other hand, the orangutan would present a silent allele *m,* which is responsible for the M-negative phenotype (which happens to be also N-negative because the N factor probably does not occur in this species). As for the gibbon, detached far earlier from the common branch, it would appear to have evolved in a fashion more or less parallel to that of man and has achieved a comparable degree of segregation between the two genes. This is demonstrated by the presence of two codominant alleles, *M* and *N,* fairly similar to those encountered in man. This type of *progressive mutation,* well known to geneticists, occurs in all taxonomic groups and involves various traits, both morphological and physiological (see below).

TABLE 6-3. Relationships among M-N genotypes and phenotypes according to Wiener's hypothesis

Genotypes	Phenotypes
MM NN	
MM Nn	MN
Mm Nn	
Mm Nn	
MM nn	M
Mm nn	
mm NN	N
mm Nn	
mm nn	O

The Genetic Model of the M-N System According to Wiener et al.

In 1972, A.S. Wiener et al. [254,270] proposed a new model to explain the inheritance of the M-N system, both in man and in the nonhuman primates. This can be diagrammed in the following way.

In all primates (including man) the M-N system is ruled by two pairs of alleles, *M/m* and *N/n*, carried on two pairs of different chromosomes. *M* and *N* ensure the synthesis of the corresponding factors and are dominant over their *m* and *n* silent alleles. Moreover, a heterostasis phenomenon occurs between M and N so that the quantity of N carried by the red blood cell diminishes when the M is present in the genotype. A species that has retained all the *M/m*, *N/n* mutations could form the combinations shown in Table 6-3.

This theory can account for all the facts observed only if one accepts that some mutations were lost during the different stages of the processes of speciation. This loss occurred by chance, which is why the distribution of the M-N factor in the main species of primates does not follow the taxonomic ladder.

In man, who shows three phenotypes—M, N, and MN—it is the *n* allele that seems to have been lost. All human beings, therefore, carry the *N* gene in the homozygous state, the variation affecting only the *M/m* pair. The relationships among human pheotypes and genotypes are shown in Table 6-4.

The hypothesis fits in well with the serological observations mentioned above (partial absorption of anti-N sera with human group M red cells; appearance of "parasite" anti-N in the sera of rabbits immunized with human M cells; more frequent occurrence of anti-M antibodies in the sera of group

TABLE 6-4. Genetics of the M-N system of man (and gibbon)
according to Wiener's hypothesis

| Genotype | Phenotype | Frequency of: | |
		Phenotype	Factors
MMNN	M	30	M = 80%
mmNN	N	20	
MmNN	MN	50	N = 70%

N individuals than in anti-N persons of group M, etc.), which suggest persuasively that all human red cells of group M contain small quantities of the N antigen.

The chimpanzees display only two phenotypes: M and MN. They have presumably lost, in the course of speciation, the *m* allele, so that all animals of this species are homozygotes *MM,* and on the phenotypic level are uniformly M-positive. Here, the polymorphism affects only the *N/n* pair, which accounts for the two phenotypes observed in this species, as Table 6-5 demonstrates.

As for gorillas, they would have lost the *n* allele, and would all, therefore, be homozygous for the *N* gene (genotype *NN*). The two phenotypes encountered in this species, N and MN, would correspond to three genotypes (see Table 6-6). We should note here that the M of gorilla would be the product of a lesser $M \rightarrow N$ mutation than the M of man, so that the *MMNN* genotype corresponds in these apes to an MN phenotype rather than an M phenotype, as is the case in man.

Actually, it seems that the gorilla has two MN phenotypes, the first (M_1N) with an M antigenic content far greater than the second (M_2N). M_1N probably corresponds to the *MMNN* genotype and M_2N to the *MnNN* genotype.

The orangutans, 50% of which carry M, always lack the N factor and would therefore have lost the *N* gene. In these species, therefore, all the animals are either M-positive or M-negative (Table 6-7.).

Finally, the gibbon would seem to have preserved three genes, *M/m* and *N;* it has lost *n* and therefore presents the same genetic model as man (as shown in Table 6-4).

In order to explain the serological relationship between M and N, Wiener et al. offered the following hypothesis: The N receptors might be located more deeply in the surface of the red blood cell than the M receptors. Thus, the factor N would be fully accessible to the antibodies when the red blood cell lacks M. When the M receptors appear, they would mask the N receptors to the point of making them almost (but not completely) inaccessible to the

TABLE 6-5. Genetics of the M-N system of chimpanzee
according to Wiener's hypothesis

| | | Frequency of: | |
Genotype	Phenotype	Phenotype	Factor
MMnn	M	60	M = 100%
MMNN } *MMNn*	MN	40	N = 40%

homozygote. This hypothesis, promulgated long before the probably biochemical association of these two factors became known, implied a connection between the genesis of N and of M, the second antigen being related to
the first as A is to H. The work of certain biochemists (Uhlenbruck [231],
Pardoe et al. [157]), and observations made by Lisowska and Kordowicz
[111] and others seemed to confirm this hypothesis: the experiments, consisting of submitting M and N antigens to a progressive enzymatic degradation,
were the first to shed light on the structure of these molecules and the
elements that lead to their formation, which can be summarized as follows.

The M and N factors are membrane glycoproteins whose terminal fraction
carries antigenic specificities.

Glycophorin, the principal erythrocytic protein, is terminated by a serine
and a threonine. This terminal first attaches an α-acetyl-galactosamine, which
results in the Tn type of polyagglutinability.

The second stage is represented by the attachment of a β-galactose, which
produces the T type of polyagglutinability, described by Friedenreich in 1930
[48]. The last two stages, according to Friedenreich, are characterized by the
binding of an N-acetyl neuraminic acid (NANA), which results in the N
factor, and the binding of a second acid in the terminal position, leading to
the M specificity.

Thus, the difference between M and N would be related to the degree of
sialylation of the molecule (i.e., the quantity of bound NANA). The N^V
specificity, defined by *Vicia graminea* lectin, and some anti-N of human or
animal origin would be represented not by the terminal part of the chain, but
by its internal fraction corresponding to the α-GalNac, β-Gal links. This is
why the N^V specificity is present on both the N and the M red blood cells
but is more effectively disguised in the second group. This could explain the
serological findings described above, according to which the M red blood
cells are capable of reacting with the anti-N antibodies (since, in this case,
the terminal part of the chain carries the N specificity, partly disguised by the
additional terminal NANA, responsible for M).

In contrast, the N antigen does not react with the anti-M antibodies, since the N factor lacks the terminal NANA portion that ensures the M specificity. This would explain why treatment of red blood cells with proteolytic enzymes suppresses their agglutinability by anti-M and anti-N antibodies.

The sequence can be represented in the following way (modified from Springer and Desai [214]):

$$-\text{O-Ser/Thr} + \alpha\text{-GalNAc} + \beta\text{-Gal} + \alpha\text{-NANA} + \beta\text{-NANA}$$
$$\text{Glycophorin} \rightarrow \quad \text{Tn} \quad \rightarrow \text{T} \quad \rightarrow \quad \text{N} \quad \rightarrow \quad \text{M}$$

Blumenfeld and Adamany [9] isolated three fragments, A, B, and C, of the human M, N, and MN glycoproteins (glycophorins). The A fragment consists of eight amino-acid residues, six of which are common to the three groups and two variable with each group (glycine and glycopeptide in A^{MM}; leucine and glutamic acid in A^{NN}; and half residues of serine, glycine, leucine, and glutamic acid in A^{MN}). Each glycoprotein carries two tetrasaccharides (2 NANA, 1 Gal and 1 GalNAc) and a trisaccharide (NANA, Gal, GalNAc) linked to a threonine and two serines. Thus, the codominant MN alleles would act on the sequence of amino-acid residues of the A peptidic fragment. According to Furthmayr [49a] the *MNSs* locus controls the synthesis of two proteins: glycophorin A (for M-N) and glycophorin B (for S-s). The two glycophorins are structurally similar and parts of the major sialoglycoproteins of the red cell membrane. The glycophorin of an M individual (glycophorin A^M) differs from that of an N individual (glycophorin A^N) by amino acids at positions 1 and 5 [110a]. A second sialoglycoprotein, glycophorin B (responsible for S and s specificities), has an amino acid sequence identical to that of glycophorin A^N in the first 23 positions and carries N activity only. S and s differ from each other by a substitution at the 29th amino acid [24a]. Biosynthesis of glycophorin A was studied by Gahmberg

TABLE 6-6. Genetics of the M-N system of gorilla according to Wiener's hypothesis

Genotype	Phenotype	Frequency of:	
		Phenotype	Factors
mmNN	N	17	M = 76%
MMNN } *MmNN* }	MN	83	N = 100%

et al. [49b] using human leukemia cell line K562; they achieved translation of glycophorin A messenger-RNA in a rabbit reticulocyte cell-free system.

Mutations Specific to Each Species

So far we have not considered qualitative differences among the mutations affecting the M-N system in various primate species. In biochemical terms, each M mutation leads to the binding on the end of the glycoprotein chain of a very large number of N-acetyl-neuraminic acid residues; the quantity of residues thus retained varies from one species to another and corresponds to the different types of M-N antigens observed.

Actually, the differences in the M factor noted among the species of primates studied so far appear to be not only quantitative but also qualitative. Along with an M antigenic pattern common to all species (a kind of "fundamental M"), and encountered in all M-positive individuals, whatever their zoological background, there exist patterns characteristic of each lineage. This is suggested by cross-reactions obtained with batteries of immune antibodies, developed by the injection of rabbits with human M red blood cells. When the reactions given by the same set of antibodies with the red blood cells of the four species of anthropoid apes and man are compared, it seems that the differences of agglutination observed among them do not occur by chance. They happen in such an order that each new mutation involves adjacent sequences or a group of neighboring sequences, over a more or less extended scope. This type of evolution has been invoked to explain certain degrees of the D^u factor that can be observed in the human species; it has been known for a long time in animals, since it was described for the first time by Dubinin in 1929 for the scale mutation of *Drosophila melanogaster.*

The transformation of N factor into M during the evolutionary process could have followed the same path and involved new sequences at each step. The qualitative differences observed by Wiener among the M antigens of the species studied could be of a stereochemical order and related to the spatial arrangement of the glycoprotein chains with M specificity in relation to those that would have retained N specificity. Nevertheless, the species of anthropoid primates now living have no direct lineal relationships. Descending at different periods from the ancestral branch, each one has followed its own evolutionary path.

This is how the dispersive (and disorderly) aspect of immunological modification came about; for example, the gibbon, isolated from the common branch earlier than the African anthropoid apes, went further in the transformation of N into M and as a result arrived at a genetic model identical to that of man.

TABLE 6-7. Genetics of the M-N system of orangutan according to Wiener's hypothesis

Genotype	Phenotype	Frequency of:	
		Phenotype	Factor
$\left.\begin{array}{l}MMnn \\ Mmnn\end{array}\right\}$	M-positive	60%	M = 60%
mmmn	N-negative	40%	N = 0%

Did this diversifying evolution occur only at the moment of the synthesis of the M factor, or had it already affected the earlier stages, at the time of the appearance of the N factor, for example? In other words, does the interspecific immunological polymorphism observed in primates for group M depend on the M gene, or is it linked to a preexisting polymorphism in the precursor stage that would already have affected the N factor? Does it correspond to any early modification or, on the contrary, to a variation introduced at the last stage and noticeable only at the end of the chain? Only the use of enough and sufficiently potent anti-N reagents will provide an answer. Unfortunately, as we have said, these anti-N antisera always had low titers and hence were of little use after the heteroagglutinins had been absorbed out. Therefore, all the investigators who have studied the N factor of primates have resorted to the use of *Vicia graminea* lectin as their principal anti-N reagent. But this reagent recognizes only very limited determinants on the glycoprotein chain, called N^V, which corresponds to the sequence β-Gal-α-GalNAc, to which the residues of neuraminic N-acetyl acids attach themselves. Because of this, its ability to identify combining sites is only approximate. The exact nature of the qualitative differences observed among the M antigens of primates will one day be elucidated by the biochemists.

A step in that direction has been reported recently by Blumenfeld et al. [9a], who found that the M-N activity of chimpanzee red cells is carried by a major membrane glycoprotein that is similar but not identical to human MM-glycoprotein (glycophorin A). The amino terminal fragment of the glycoprotein molecule was found to be a glycooctapeptide whose amino acid composition and partial sequence indicated that it is an intermediate form of the human M- and N-glycooctapeptides. Structural similarities of this fragment to the human antigens strongly suggest that the amino terminus bears the major antigenic determinant(s) of the molecule and that the occurrence in this region of numerous, albeit rare, variants among humans and in chimpanzees indicates that the corresponding coding sequence of the structural gene is particularly susceptible to mutational events. From the observations made by Blumenfeld and collaborators it may be concluded that such a structure as the one observed in chimpanzees could be a precursor of the human M and N glycoproteins, since single base mutations leading to the substitution either of serine at position 1 by leucine, or of glutamic acid at position 5 by glycine, would give rise to the N and M human glycoproteins, respectively.

The Current Genetic Model of the M-N system

How do we picture a model that now reconciles the immunological and biochemical data and is valid as an explanation of the findings in man as well as in nonhuman primates?

Two outlines are theoretically possible.

1. The model that was the first to be proposed introduced a polyallelic series, such as exists in many animals and plants. See, for example, the *w* (white) mutation[1] that decolorizes the eye of the *Drosophila,* or the *v* (vestigial) mutation that reduces the size of its wings to the point of disappearance; the two are part of a progressive series of at least twelve mutations for the first locus and eight for the second.

In some way, we can imagine a series of progressive mutations, going from anthropoid apes to man, each one of which is capable of transforming a larger quantity of N substance into M. But this series, instead of occurring completely in one species, is more or less dispersed in the phylogenic scale.

This transformation is less advanced in African anthropoids (such as the gorilla and the chimpanzee), which carry a gene capable of ensuring the simultaneous synthesis of M and N, than in the Asiatic anthropoids (orangutan and gibbon), which, together with man, possess an *M* allele capable of inducing the transformation of most of N into M.

2. A second model, which the majority of investigators now accept, and which is directly inspired by the hypothesis proposed by Wiener et al., implies that the M-N system is controlled by two pairs of independent alleles, *N/n* and *M/m*, each acting at a different moment of biosynthesis (as *H* and *A,* for example). In these two pairs, only the first expression (*N* and *M*) is active; the second is a silent gene (*n* and *m*). The *N* mutation ensures the

---

[1]The white mutation, which removes all pigment from the eye, has a series of progressive mutations that increasingly decolorize the ommatidia of the *Drosophila.* These are *wild, coral, blood, eosin, cherry, apricot, honey, buff, tinged, ivory, ecru,* and *white.* In the same way, the series of mutations that reduce the size of the wings ranges from the normal type to the apterous type with progressive reductions, whose principal terms are *normal, nicked, nick, notched, antlered, strap, vestigial,* and *no-wing,* to name only the most common.

transformation of the polyagglutinability T antigen into the N antigen, and the *M* mutation is responsible for the change of N into M (in each case, by the attachment of an α-NANA residue).

THE HENSHAW (He) AND HUNTER (Hu) ANTIGENS

The Henshaw antigen (He) was discovered by Ikin and Mourant [67a] in 1951, when they were immunizing rabbits with human M red blood cells for the production of anti-M antibodies. They obtained a new reagent capable of recognizing a hitherto undescribed factor present in blacks (with frequencies that are scarcely 5%) and absent in the white and yellow races. They called it the Henshaw factor, or He. This antigen belongs to the M-N system, but its association with different factors depends on the ethnic group considered. It is found to be associated predominantly with *NS* in New York blacks (3.2% of He+ subjects), with *Ns* in Papua, with *MS* in the Congolese black population and the Hottentots, and with *Ms* in the inhabitants of the Cape Providence (see Race and Sanger [159]).

The Henshaw factor was present in the majority of gorillas studied by Wiener, et al. [270] (two negative subjects of fifteen). However, it has not yet been found in chimpanzees or in the other apes.

Another factor, Hunter (Hu), was discovered by the antibody produced by Landsteiner et al. [96a] in 1934 in a rabbit immunized with the red blood cells of a U.S. black. This factor is present in blacks (up to 25%) but is very rare in whites. Just as is the preceding factor, it is associated with the M-N-S system. Its immunological definition is otherwise far more blurred than that of the preceding factors, to which it nevertheless seems to be closely related. The rarity and low avidity of usable reagents has kept it from being investigated in nonhuman primates.

It is probable that *He* and *Hu* mutations are situated on loci very close to the one that carries *M* or *N*. In families studied by Shapiro [192], the He factor always segragated, depending on the case, with Ms or Ns or MS, without detectable crossing-over.

At present, it is not known whether *He* and *Hu* occupy the same locus, i.e., whether they are true alleles. If this were the case, they would belong to a series composed of three mutations: *He, Hu,* and *h,* the first two being responsible for the synthesis of the corresponding factors and the latter recessive, with no discernible action. In the white and yellow races, *h* could be considered a public factor without known phenotypic expression.

Nevertheless, it is also possible that *He* and *Hu* are not true alleles but occupy adjacent loci. Each one would then represent the dominant expression of a pair of alleles, *Hu/hu* and *He/he,* the two recessive expressions *hu* and *he* behaving as silent genes. A study of a black family by Rosenfield et al. [165] suggests a possible transmission of NS, Hu, He complex. This could constitute an argument in favor of the nonallelism of *He* and *Hu,* although the illegitimacy of a child could not be excluded.

Based on the reactions obtained with a serum produced by immunizing a rabbit with type N+He+ human red cells, Wiener and Rosenfield [277] postulated the existence of a factor called Me that would combine the specificities M and He. Their findings confirmed the earlier findings of Ikin and Mourant [67a], who discovered immune antibodies that were capable of detecting the factor He but somewhat associated with anti-M, from which it could not be separated by absorption. But the genetic interpretations offered by the two teams were different. While Mourant and Ikin attributed the specificities M and He present in their serum to the action of two autonomic antigens (not unlike the model of the Rh system proposed by Fisher and Race), Wiener and Rosenfield [277] postulated the action of a single gene that belonged to the same polyallelic series as M. The gene, which they called *Me,* is presumed to be an intermediary between *M* and *He* and to control the simultaneous appearance of both specificities, M and He.

ANTIGENS OF THE MILTENBERGER SERIES

This is a third series of factors with a very low frequency, which are clearly antigenic for man and are closely associated with M-N. The first antibody, corresponding to one of them, was discovered by Levine et al. [108a] in a patient suffering from hemolytic anemia (Mrs. Miltenberger, whence came the name of the antibody, the abbreviation anti-Mi^a). Later, the anti-Verweyst (anti-Vw), the anti-Graydon (anti-Gr), the anti-Murrell (anti-Mu), and the anti-Hill (anti-Hil), etc., were discovered.

The red cells bearing the antigens of the system are agglutinated by one or several of these antibodies, i.e., they can carry several specificities. According to the way in which these antigens are combined in one individual, Cleghorn [23] divided them into five classes (I to V). All these factors are very rare in whites (except in some very endogamous isolates). They always segregate with the M or N factor, to which they are linked.

Later, a good many other antibodies were discovered that turned out to be satellite antigens of the M-N system (Vr, Ri(a), St(a), Mt(a), Cl(a), Ny(a), Sul, Far, etc.). Their incidence is always very low.

During the course of some experimental alloimmunizations between chimpanzees, one of our animals produced an anti-V^c antibody (which will be discussed later) that specifically reacted with human red blood of the Miltenberger-positive (Mi^a-positive) type but not with other human red cells [200]. This demonstrates that the Mi^a factor exists in the chimpanzee and that it must be, as in man, linked to the M-N system, since, as will be shown later, the V^c belongs to a chimpanzee blood group system that is a counterpart of the human M-N system.

ANTIGENS OF THE S/s/U SERIES

This series of factors is at present known to occur only in man. It is possible, however, that the absence of these specificities from red cells of nonhuman primates is only apparent; it may be caused by the weakness of the typing reagents. After absorption of the heterospecific agglutinins present in the human serum, the residual titer of anti-S or anti-s becomes too weak to give meaningful results with the blood of primates. It may be, however, that antigens of this series are exclusive attributes of the human lineage.

This series is made up of at least three alleles: *S, s, Su*. The first two were discovered thanks to irregular antibodies present in the sera of women who had had incompatible pregnancies and in patients who had suffered transfusion reactions. The S was discovered first by Walsh and Montgomery in 1947 [237]; the s was discovered in 1951 by Levine et al. [106]. These two factors are controlled by two codominant genes occupying a locus situated close to the M-N series, with which they show an intimate relationship. Unlike the M and N factors, which are usually not immunogenic for man, perhaps because of their close structural similarity, S and s are clearly antigenic and can lead to serious incompatibility accidents, as is shown by the circumstances of their discovery.

A recessive allele, *Su*, observed so far only among blacks, appears to be devoid of any antigenic expression. The homozygous individuals *Su/Su* are phenotypically S−s−. Nevertheless, in 1953 Wiener et al. [290] described an antibody anti-U with peculiar properties: 1) It agglutinated all red cells, either S *or* s (phenotypes S+s−, S−s+, or S+s+). 2) It also gave positive reactions with 16% of the S−s− individuals, while the blood of the remaining 84% of S−s− persons was not agglutinated. Antibody anti-U is not, therefore, a mixture of anti-S and anti-s antibodies, nor a cross-reactive anti-S+s (as was previously believed). 3). All U-negative individuals tested were either type N or type MN, thus indicating a relationship between U and the M-N system.

Crossing-over between the M-N and S-s loci has been unquestionably observed in several families, although its occurrence is too rare to define a rate of recombination [50]. This demonstrates that the two corresponding chromosome sequences are definitely independent. They are, however, related functionally, as shown by the fact that erythrocytes of type MS−s− are totally lacking in N antigen, both in U-positive and U-negative subjects [1].

ANTIGENS OF THE $V^c/A^c/B^c/D^c$ SERIES

This series of antigenic properties, which constitute a part of the M-N blood group system, does not occur in man but appears on the red cells of anthropoid apes, particularly chimpanzees *(Pan troglodytes* and *P. paniscus),* in which it was originally described [267].

The series comprises four factors:

1) Factor V^c, recognized by antisera produced in chimpanzees either by isoimmunization (injection of a V^c-negative chimpanzee with V^c-positive chimpanzee red cells) or by heteroimmunization of a V^c-negative chimpanzee with human red cells (of either the M or N type) [127].

Subsequently, it appeared that these anti-V^c reagents gave more or less the same reaction with chimpanzee red cells as did the anti-N^V extract from *Vicia graminea.*[2] It may therefore be concluded that the anti-N^V of plant origin and the anti-V^c obtained by alloimmunization or heteroimmunization of the chimpanzee recognize somewhat related combining sites of the antigenic molecule. The N^V factor, precursor of N, is present on all human red blood cells (but more accessible in group N than in group M) and also on the red cells of some anthropoid apes, namely those of the V^c-positive type.

Red cells of all lowland gorillas *(Gorilla gorilla gorilla),* as well as all mountain gorillas *(G. gorilla beringei)* studied until now, are agglutinated by all anti-V^c reagents. But considering the small number of animals tested, it is difficult to know if this reflects the true monomorphism of gorillas with respect to the V^c factor or is rather due to the accidental composition of the sample.

2) Factors A^c, B^c, D^c. These chimpanzee red cell specificities were discovered as the result of a series of experimental alloimmunizations carried out by Wiener et al. [267] in which several blood factors were defined and were named after the consecutive letters of the alphabet, A^c, B^c, D^c, E^c, and F^c (in each case the exponent "c"—for chimpanzee—as explained earlier, was used as a reminder that there were specific factors that had nothing to do

---

[2]Parallel reactions obtained with anti-V^c and anti-N^{Vg} reagents prompted Wiener to assign to the latter the symbol "anti-N^V."

TABLE 6-8. Alleles of the V-A-B-D system and
their presumed action

V	synthesis of V^c	
v^A	synthesis of A^c	codominant
v^B	synthesis of B^c	
v^D	synthesis of D^c	
v	silent allele	recessive

with human A-B-O groups or C-D-E specificities of the Rh system). While the factors C^c, E^c, and F^c were later found to belong to the C-E-F- system, a chimpanzee counterpart of the Rh system (to be discussed in the next chapter), the A^c, B^c, and D^c specificities were, from the beginning, associated with the M-N system because of their serological properties. (While the factors A^c, B^c, D^c, and V^c are destroyed by the action of proteolytic enzymes (as are the M and N factors), the other three specificities—C^c, E^c, and F^c— are best identified by enzyme-treated red cell techniques, as are the factors of the Rh system.) The study of the distribution of these three factors in 245 unrelated chimpanzees, as well as analysis of their inheritance in several chimpanzee families, confirmed that these factors were genetically related to V^c and, therefore, linked to the M-N system [258].

To explain the inheritance of the V^c, A^c, B^c, and D^c factors, Wiener et al. [255] proposed a model that assumed the existence of five alleles: four codominant alleles—V, v^A, v^B, and v^D—that are responsible for the synthesis of the corresponding red cell factors and a recessive allele, v, that appears to be a silent gene. The relationships among these genes and their end-products are shown in Table 6-8.

The five alleles assumed to exist contribute to the eleven phenotypes theoretically expected, of which all but one were actually observed. The missing type, v.O, corresponds to the homozygous genotype vv (see Table 6-9).

ANTIGEN W^c

We should recall that an anti-V^c reagent, obtained from the serum of a V^c-negative chimpanzee immunized with V^c-positive chimpanzee red cells, was found to react specifically with human red cells of Mi^a (Miltenberger-positive) type. Although originally thought to be part of anti-V^c, the antibodies reactive with Mi^a red cells were subsequently found to be different from anti-V^c. A few chimpanzees were recently discovered of the unusual type V.AB (their red cells gave positive reactions with anti-V^c, anti-A^c, and anti-B^c), in apparent defiance of the proposed genetic model of the V-A-B-D system (Table 6-9). By absorbing the anti-V^c serum with red cells of those

TABLE 6-9. Serology and genetics of the chimpanzee V-A-B-D blood group system

Designation	Reaction with serum of specificity				Number		Possible genotypes
	Anti-V^c	Anti-A^c	Anti-B^c	Anti-D^c	Observed	Expected	
v.O	−	−	−	−	0	0.45	vv
v.A	−	+	−	−	43	43.00	$v^A v^A$, or $v^A v$
v.B	−	−	+	−	27	27.00	$v^B v^B$, or $v^b v$
v.D	−	−	−	+	5	3.20	$v^D v^D$, or $v^D v$
v.AB	−	+	+	−	49	53.30	$v^A v^B$
v.AD	−	+	−	+	9	13.20	$v^A v^D$
v.BD	−	−	+	+	12	10.10	$v^B v^D$
V.O	+	−	−	−	17	17.0	VV, or Vv
V.A	+	+	−	−	47	40.6	Vv^A
V.B	+	−	+	−	27	31.2	Vv^B
V.D	+	−	−	+	9	7.7	Vv^D

Estimated gene frequencies:
$v = 0.0429$
$v^A = 0.3761$
$v^B = 0.2891$
$v^D = 0.0715$
$V = 0.2206$ $\chi^2_{(6)} = 5.310, 0.50 < P < 0.70$

rare chimpanzees, it could be established that, in fact, the anti-V^c reagent contained a second, slightly weaker antibody, tentatively called anti-W^c, that recognized a "new" specificity, W^c, on chimpanzee red cells. The W^c is the chimpanzee counterpart of the human Mi^a, as shown by the fact that the anti-W^c component can be removed from anti-$V^c + W^c$ serum by absorption either with chimpanzee V^c-negative, W^c-positive red cells or with human Mi^a-positive blood. When the anti-V^c serum thus absorbed was used side by side with unabsorbed anti-V^c reagents for retesting the blood of the chimpanzees of the alleged V.AB type, they were all found to be type v.AB, W^c-positive.

PHYLOGENY OF THE M-N BLOOD GROUP SYSTEM

The first biochemical evidence of the intimate closeness of the V-A-B-D and W^c specific sites and the M-N combining sites on the membrane of chimpanzee red cells was provided recently when M-N-active glycoprotein preparations from chimpanzee red cell membranes were made available (courtesy of Professor Olga Blumenfeld, Albert Einstein College of Medicine) for serological testing. The direct inhibition and antiglobulin inhibition tests performed with preparations from red cells of four chimpanzees of various V-A-B-D and W^c types showed (Table 6-10) that these specificities

TABLE 6-10. Inhibition tests with glycoprotein preparations obtained from the red cells of chimpanzees of various V-A-B-D and W^c types

Antiserum vs. indicator red cells	Inhibition titer* of the glycoprotein preparations from chimpanzee red cells				
	Chi-91, Vito A_1,M, v.B,w^c	Ch-21, Melilot A_1,M, v.AB,W^c	Ch-85, Billy $A_{1,2}$,MN, V^q.B,w^c	Ch-151, Ira A_1,MN, V.A.	Saline control
1. Anti-A (human) vs. A_2 (human)	0	0	0	0	0
2. Anti-N (rabbit) vs. N (human)	0	0	4	1	0
3. Anti-N^V *(Vicia unijuga)* vs. N (human)	nd	nd	Over 256	Over 256	0
4. Anti-N^V *(Vicia unijuga)* vs. M (human)	nd	nd	128	32	0
5. Anti-M (rabbit anti-chimp.) vs. M (human)	Over 256	Over 256	Over 256	Over 256	0
6. Anti-M (rabbit anti-chimp.) vs. M (chimpanzee)	32	32	32	16	0
7. Anti-V_i^c (chimp. isoimmune) vs. V^c (chimpanzee)	0	0	0	96	0
8. Anti-V^c+V^q (chimp. isoimmune) vs. V^q (chimp.)	0	0	64	Over 256	0
9. Anti-V^c+W^c (chimp. isoimmune vs. W^c (chimp.)	0	64	0	Over 256	0
10. Anti-V^c+W^c** (chimp. isoimmune) vs. Mi^a (human)	0	16	0	256	0
11. Anti-A^c (chimp. isoimmune) vs. A^c (chimp.)	0	64	0	64	0
12. Anti-B^c (chimp. isoimmune) vs. B^c (chimp.)	96	96	96	0	0

nd = not done.
*Titer expressed in units equal to reciprocal of the highest dilution of preparation that inhibits the respective serum.
**Absorbed with human B red blood cells.

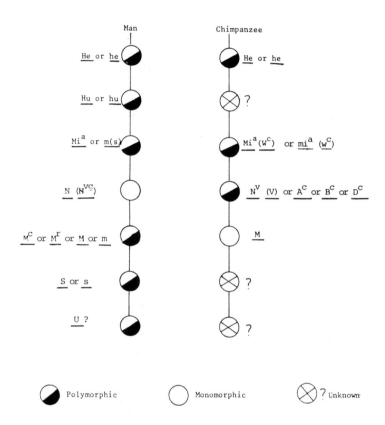

Fig. 6-1. Hypothetical comparison of the chromosomal segment of the M-N loci of man and of chimpanzee (the order of the loci is purely arbitrary).

are carried with the same glycoprotein fraction as the M and N specificities. These two classes of specificities, however, seem to be serologically independent from each other. Further biochemical fractionations are necessary to identify the chemical structures responsible for either M-N or V-A-B-D and W^c specific reactions.

When considering the various appearances of the M-N blood groups in higher primates, in particular the anthropoid apes, which are the closest to man, one must be struck by the exceptional complexity of this blood group system. The system is constructed around one basic element that is identical or very similar to the N^V factor, the precursor of the N and M factors, present (though not always serologically detectable) on all human red cells. It has also been found on the red cells of anthropoid apes, where it was called V^c. It was detected in all gorillas so far tested (which were also all N-positive), but it is polymorphic in chimpanzees, which can be either V^c-positive or V^c-negative (all V^c-positive animals are also N-positive, while the V^c-negative type corresponds to an N-negative reaction of the red cells). Obviously, the existence of V^c-negative chimpanzees enabled production of the first anti-V^c reagents by means of alloimmunization.

Around this central pattern, common to several zoological groups, which constitutes a veritable immunological bridge among species and doubtless characterizes a certain taxonomic level, some secondary patterns attached themselves: here belong factors He and Mi^a (or W^c), which appeared sufficiently early during evolution to be encountered in several lineages. They correspond to *paleosequences.*

Later, some other patterns emerged, perhaps as a result of the duplication of the same chromosome segment; their delayed appearance would explain why their occurrence is limited to a single species, of which they are probably contemporaries. These *neosequences* are represented by the S/s pair of man and by the $A^c/B^c/D^c$ series of chimpanzee. There are perhaps other, as yet undescribed, comparable factors in related species of apes. The study of those still unknown antigenic factors may require alloimmunization experiments in species that are rare and difficult to handle in captivity.

The series of factors $A^c/B^c/D^c$, peculiar to chimpanzees and connected with the M-N system serologically, resemble the human S/s factors: they are likely to cause alloimmunization, they are destroyed by ficination, etc. It might be concluded that there is a kind of phylogenic connection between the two series and that each one derives from an identical chromosomal segment inherited from a common ancestor that has evolved in a special way in each species. This hypothesis, however, is barely plausible: whereas V^c is immunologically almost identical to N^V and genetically constitutes an allele of A^c, B^c, D^c, the S/s mutations are linked to the MN locus but are not true alleles, as shown by rare crossings-over observed between the two loci. These relationships can be demonstrated by the outline given in Figure 6-1, which does not prejudge in any way the respective positions of the different loci.

Thus, on the MN chromosome 1) an original segment seems to exist that in man carries the S/s/U locus and that remains silent or is nonexistent in anthropoid primates, and 2) a common segment exists that carries either mutations found in man as well as in apes—M-N, Mi^a, N^V, and V^c—or mutations belonging only to anthropoids—A^c, B^c, D^c. Both kinds of mutations may be under the pressure of natural selection [216].

Blood Groups of Primates, pages 75–90
© 1983 Alan R. Liss, Inc., 150 Fifth Avenue, New York, NY 10011

7

The Rhesus System

History of the Discovery of the Rh System 75
The Rhesus System in Man 76
The Rhesus System in Nonhuman Primates 78
    Rh Factors Recognized by Reagents of Human Origin 78
    Rh Factors Recognized by Reagents of Simian Origin: The $C^c/c^c/E^c/F^c$
    Series 81
    Genetic Model of the R-C-E-F System of Chimpanzee 84
    The R-C-E-F System in Other Nonhuman Primate Species 88
The Rhesus System in Evolution 89

HISTORY OF THE DISCOVERY OF THE Rh SYSTEM

The discovery of the Rhesus system in 1940 by K. Landsteiner and A.S. Wiener [98] was certainly the most important event since the description of the basic A-B-O blood groups by this same Landsteiner, at the beginning of the century. It is this discovery that shed light on the process of fetomaternal alloimmunization, explained transfusion reactions, whose causes had until then remained unknown, and led to the description of a number of important "new" genetic blood systems. Two sciences were about to be launched: immunogenetics and the immunology of evolution.

We have seen (in Chapter 4) the fundamental consequences that these discoveries had in the fields of anthropology, zoology, and evolution, and how the substitution of populational thinking for typological thinking was perhaps the most important conceptual revolution in the life sciences during this century.

These discoveries were not actually due to chance, as sometimes happens in biology, but were the result of a skillfully conceived plan. We have already described how Landsteiner and Levine discovered the M-N factors by immunizing rabbits with human red blood cells. Later, during research into new variants of M and N antigens, Landsteiner and Wiener used immune sera prepared by injecting rabbits, not with human red blood cells, but with red cells from various species of monkeys. In particular, by injecting rabbits with rhesus monkey *(Macaca mulatta)* red cells, the two authors obtained a specific anti-M along with other antibodies.

One of these antibodies, isolated by specific absorption, was found to agglutinate the red cells of 85% of U.S. whites (who were called Rhesus-positive: Rh+); the 15% whose red cells were not agglutinated were called Rhesus-negative: Rh−. Very soon it became clear that the so-called Rhesus factor was responsible for certain serious cases of icterus in the newborn that were ascribed to an immunological incompatibility between the mother and fetus, as Levine and Stetson had already suspected in 1939 [108]. This factor was also found to be at the root of unexplained transfusion reactions that sometimes occurred during isogroup transfusions.

THE RHESUS SYSTEM IN MAN

When the irregular antibodies occurring in polytransfused subjects or in pregnant women showing alloimmunization reactions were studied, a series of specificities linked to the initial Rhesus factor (which became standard Rhesus: D or Rh_0) were discovered. These specificities were given two distinct nomenclatures. The first one, by A.S. Wiener, stems from what was called the unilocular theory. It postulated the existence of a single polyallelic series with each gene able to ensure the synthesis of two or three factors. The second (triocular theory), by R.A. Fisher quoted by Race [158a], assumed the existence of three pairs of closely linked alleles on the same chromosome. Table 7-1 summarizes the relation between the two nomenclatures.

Little by little, Fisher and Race's nomenclature was generally adopted; it is considered more convenient to handle and easier to remember. In addition, it has the advantage of simplifying the classification of all newly discovered

TABLE 7-1. Comparison of two nomenclatures for the Rh blood group system

Wiener (a series of alleles)	Fisher and Race (three pairs of alleles)
R^1	CDe
R^2	cDE
R^o	cDe
R^z	CDE
R'	Cde
r''	cdE
R^y	CdE
r	cde

mutations at one of the three loci: At the D/d locus the mutations include D^u and D^w. D^u does not correspond to a single type but includes several graded variants—from the highest, which corresponds to the kind of red cells that are agglutinated by almost all anti-D (Rh_0) reagents, down to the lowest degree, revealed only by antiglobulin technique. Family studies demonstrated that, generally, the same degree of D^u is encountered in the same stock; each degree must therefore correspond to an autonomous mutation. The number of these mutations must be very high, since the number of D^u degrees thus far described, is itself very high.

The two other series are C^w, C^x, C^v, C^u, c^g at the C/c locus and E^w, E^T, e^s at the E/e locus.

Race and Fisher's model expresses clearly the deficiency phenomena affecting either the E/e locus: type C^wD and $cD-$ chromosomes; or two loci: "-D-" type chromosome; or even the three loci: Rh_{null} type, but this type can have various genetic causes.

However, the oversimplification of this model causes it to be far from immunological reality. Reducing a collection as complex as the Rhesus system to three pairs of alleles that are put on an equal footing is certainly not very plausible. Let's recall, among other circumstances, the following:

1. The D (Rh_0) factor occupies a central, privileged position, as we shall see when studying the Rhesus system of the nonhuman primates.

2. There are functional relationships between the two series postulated by Race and Fisher (C/c and E/e) that give rise to the compound antigens, products of the activity of two neighboring loci situated in *cis* position—e.g., Ce (rh_i), ce^s (hr^V); CE (rh), ce (hr).

3. G antigen, present on all rhesus chromosomes with the exception of the cde (r) and cdE (r'') types, is not explicable by the existence of the three independent loci.

4. The D (Rh_0) antigen is not homogeneous but is composed of a mosaic of factors, the first of which (Rh^A, Rh^B, Rh^C, Rh^D) were discovered by A.S. Wiener. They defined the partial Rh_0 antigens (cognates of Rh_0) that can, according to Tippett [229a], be classified into at least six categories.

5. Finally, no crossing-over has yet been directly observed despite the tens of thousands of families studied throughout the world (whereas it has been observed for MN Ss).

Actually, the discoveries of molecular genetics, in particular the fine structure of the gene, by now synonymous with cistron, and the regulator systems that intervene in it, today make the above discussion out of date. New views of the immunogenetic blood factors are now being considered— for example, the one proposed in 1973 by Rosenfield et al. [164] (see also R. Sanger [190]).

At the present time, the following seems to be the case in man:

1. A fundamental antigen exists, D (Rh_0), that plays the most important role on the genetic and immunological levels of the Rh system. This antigen, which previously was believed to be homogeneous, is actually composed of a mosaic of factors.

2. Besides this antigen, two other series of antigens exist that can each show a certain number of variations. These variations correspond to two genetically independent mutations, transmittable in infinitesimal detail.

3. These three series of factors do not correspond to units that are independent of each other; they give rise to numerous interactions among themselves. In particular, it seems that some loci shelter the genes that not only are responsible for the synthesis of a structural factor, but also play the role of a regulator gene for the neighboring loci.

THE RHESUS SYSTEM IN NONHUMAN PRIMATES

A system as complex as the Rhesus system certainly did not appear all at once at the time of hominization (speciation leading to man). It had to be "prepared" a long time in advance during the course of evolution and had to exist, at least in its initial state, in the preceding evolutionary stages.

The fact that the Rh system had first been discovered at the time of immunological experiments involving the injection of *Macaca mulatta* red blood cells into rabbits must have inspired immunologists to search for structures in infrahuman primate erythrocytes that might be linked to the Rhesus system. Such studies revealed the extraordinary complexity of these immunological structures, which on the phylogenic level have not yet divulged all their secrets.

At present, it is only in the anthropoid apes, which are the closest to man, that sequences linked to the Rhesus system of man have been identified [273]. As far as these sequences are concerned, one must distinguish two immunological groups (each one possessing a particular phylogenic significance):

1. The first contains factors that have, at least initially, been recognized by human reagents (anti-Rh_0, anti-hr'). They are therefore antigens common to man and certain anthropoids.

2. The second has factors that are limited to one lineage or to a group of very similar stocks (E/e in man, C^c, E^c, F^c in African anthropoids.

We may therefore discuss these two series of factors in succession.

Rh Factors Recognized by Reagents of Human Origin

The initial study of the blood groups of the great anthropoid apes was carried out using antisera of human origin, the only ones available at the time. But this presented some difficult technical problems. Human sera always contain anti-primate heteroagglutinins that need to be eliminated in order to obtain sufficiently specific reagents. For this, absorptions or dilutions are carried out. However, the residual specific antibody must be strong enough to be suitable for typing primate red cells. The immunologist is, therefore, forced to maneuver between narrow limits, beneath which the antibody is not specific and above which it is too weak to give interpretable results.

The first studies were conducted in 1953 by Wiener et al. [248] and similar studies have been continued since then. The observed results can be summarized as follows:

Anti-D (anti-Rh$_0$) Antibody. The reactions obtained by a *nonabsorbed anti-D* serum are always positive against chimpanzee and gorilla red blood cells and negative against gibbon cells. The results are uninterpretable in the orangutan because of very strong anti-orang heteroantibodies that are regularly found in human serum; their absorption leaves behind no trace of agglutinating activity.[1] On the other hand, the human *anti-D properly absorbed* by the red blood cells of certain chimpanzees allows the demonstration of an immunological polymorphism; indeed, the red blood cells of certain animals are agglutinated by this reagent while others are not.

Anti-R^c antibodies, initially called anti-L^c by Wiener et al. [292a], identical to those found in human absorbed anti-D sera, were obtained by alloimmunization of the chimpanzee (injection of R^c-positive red blood cells into an R^c-negative animal). The antibodies defined the R^c factor, which allows the division of the chimpanzees into R^c-positive and R^c-negative types. The factor R^c was found to be intimately related to the human D (Rh$_0$), with which it shares a good number of specificities. In fact, the anti-R^c serum, whether it is prepared by alloimmunization of an R^c-negative chimpanzee by R^c-positive red blood cells or comes from a human anti-D antibody properly absorbed by the R^c-negative chimpanzee red blood cells, can agglutinate all ape R^c-positive red blood cells and also all human Rh-positive (Rh$_0$, or D) red blood cells [199].

---

[1]Only at the time of this writing did we come across a pregnant orangutan whose red cells, when used for absorption of the human anti-Rh$_0$ reagents, were able to fractionate the antiserum in such a way that it agglutinated some but not all of orangutan red cells, thus probably differentiating between Rh-positive and Rh-negative individuals. Since the Rh-negative type appears to be quite rare among orangutans, it is possible that the red cells previously used for absorptions of human anti-Rh reagents were all of the Rh-positive type, and they thus removed from the antisera not only the nonspecific agglutinins but the type-specific anti-Rh$_0$ antibodies as well.

Conversely, if the human anti-D (anti-Rh_0) serum is absorbed by R^c-positive chimpanzee cells, the anti-R^c is removed, but an anti-Rh_0 fraction remains reactive with all human Rh-positive red blood cells, although to a lower than original titer. Results of an actual experiment that involved comparative titrations and absorptions of chimpanzee isoimmune anti-R^c serum and human anti-Rh_0 (D) reagent, using chimpanzee and human red cells of various types, are given in Table 7-2.

The R^c factor of African anthropoid apes and the human Rh_0 possess the same antigenic fraction that must have been present in the common ancestor and that has been preserved, without modification, after the separation of the two lineages. This fraction could be called simply R, without any exponent. Onto this common fraction, patterns attached themselves that belonged to each of the two phyla. The latter fractions are identified by residual reactions that remain either after absorption of human anti-D serum with chimpanzee R^c-positive red cells, or after absorption of chimpanzee anti-R^c serum with human Rh-positive (Rh_0-positive) red cells.

The two-way evolution of the precursor R can be represented by the following diagram:

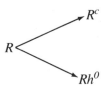

We may recall that in 1967, Masouredis et al. [119], by eluting the anti-Rh_0 coated on the human red blood cells, obtained an anti-D reagent that revealed a polymorphism in the chimpanzee. It is probable that this antibody recognized, among others, the R^c factor, with which we are here concerned.

The analysis of the transmission of R^c demonstrates that this factor is expressed in the homozygote as well as in the heterozygote. It is therefore transmitted as a dominant character. Recently, a variant has been described that gives reactions intermediate between R^c-positive and R^c-negative, with all the anti-R^c antisera. This form, which would correspond to the human D^u mutation, was initially designated by Germanic R with supercript c; it seems preferable to us, for the sake of uniformity, to call it R^{cu} (or by abbreviation R^u).

The study of three families of chimpanzees in which R^{cu} was encountered demonstrated that this mutation was transmitted as a factor dominant over r^c (and probably recessive to R^c).

Anti-C (anti-rh') and anti-c (anti-hr') antibodies. The red blood cells of all the African anthropoid apes as well as those of the gibbon are agglutinated by the anti-c (anti-hr'). However, its presence on the red blood cells of orangutans is difficult to ascertain because of very strong heterospecific antibodies in the reagents that make the interpretation of the reactions very doubtful. On the other hand, no ape blood is agglutinated by the anti-C (anti-rh'). This would seem to demonstrate that the chimpanzee, the gorilla, and the gibbon possess an antigen identical to the human c (hr') factor. However, no polymorphism is observed; none of the animals studied carry the C (rh') allele, which is, as we know, very frequent in man.

Anti-E (anti-rh'') and anti-e (anti-hr'') antibodies. These antibodies do not react with the red blood cells of any of the anthropoid apes tested so far. The chromosome segment responsible for the synthesis of the E (rh'') and the e (hr'') factors is apparently not expressed in anthropoid apes.

It is possible that this segment does not occur in this taxonomic group but emerged as the result of chromosomal rearrangements that accompanied the processes of hominization and consisted of duplication of certain chromosomal segments. In that case one could think, for example, that the segment responsible for c (hr') in anthropoid apes was duplicated to create the E/e segment of man. There is, however, another possibility, i.e. that this segment does exist in apes but remains suppressed, or that it corresponds to another series of alleles, specific for the African apes, namely the C-E-F series, which is discussed below.

Rh Factors Recognized by Reagents of Simian Origin: The $C^c/c^c/E^c/F^c$ Series

We have seen in the preceding chapter how Wiener et al. discovered a certain number of factors of chimpanzee red cells that are linked to the M-N system (V^c, A^c, B^c, D^c) by means of antibodies obtained by alloimmunizations among randomly selected chimpanzees. During this series of experiments, some antibodies were obtained that recognized factors independent of the M-N system; some were shown to be related to the Rhesus system. These are $C^c/c^c/E^c/F^c$ (the exponent is meant to identify the antigen specificity described for the first time in the chimpanzee and to differentiate it from human antigens assigned the same symbol).

In this group, the c^c factor is of particular interest. In fact, it is linked to

TABLE 7-2. Results of comparative titration and absorption experiments with chimpanzee anti-R^c and human anti-Rh$_0$ sera

Tested with red cells of:	Titer[a] by ficin method of chimpanzee anti-R^c serum after absorption with red cells					Titer[a] by ficinated red cell method of human anti-Rh$_0$ (D) serum after absorption with red cells				
	Unabsorbed	Pooled chimpanzee R^c-positive	Chimpanzee No. 169 (type R^u)	Pooled human Rh-negative	Pooled human Rh-positive	Unabsorbed	Pooled chimpanzee R^c-negative	Pooled chimpanzee R^c-positive	Chimpanzee No. 169 (type R^u)	Pooled human Rh-positive
CHIMPANZEE (R-C-E-F TYPE)[b]										
No. 17 Mack (Rc$_1$)	96	0	24	48	12	640	40	0	40	0
No. 256 Berry (Rc$_1$)	96	0	24	48	12	960	40	0	40	0
No. 261 Champ (RCFc$_2$)	96	0	22	48	12	960	40	0	40	0
No. 263 King (RCEc$_2$)	96	0	24	48	14	640	30	0	30	0
No. 169 Possum (R^{u}CEF)	6	½	0	½	½	240	0	0	0	0

No. 240 Stan (rc_1)	0	20	0	0	500	0	0	0	0
No. 643 Sean (rc_1)	0	20	0	0	500	0	0	0	0
HUMAN (Rh TYPE)									
Roman F. (Rh_1rh)	0	240	120	120	960	0	24	0	48
Frank S. (Rh_1Rh_2)	0	240	120	120	960	0	20	0	48
M.C. (Rh_2Rh_2)	0	320	240	240	960	0	16	0	64
1382-4 (Rh_0)	0	320	100	100	900	0	12	0	48
1382-5 (rh'rh)	0	0	0	0	0	0	½	0	½
James S. (rh"rh)	0	0	0	0	0	0	½	0	½
1382-8 (rh)	0	0	0	0	0	0	0	0	0

[a]The titer expressed in units is the reciprocal of the highest dilution of the reagent that gives a distinct (one-plus) reaction. Where no tube showed a one-plus reaction, the endpoint was estimated by interpolation.

[b]For explanation of the R-C-E-F types, see text on following pages.

R^c as in man C is linked to D: i.e., it is frequently (but not always) inherited together with it. Furthermore, the human anti-D (anti-Rh_0) reagents often possess, in addition to the anti-R^c specificity, an anti-c^c specificity (just as there are human anti-CD sera). This has been confirmed by fractionation absorptions.

An anti-Rh_0 absorbed by chimpanzee R^c-positive, c^c-negative (i.e., RC type) red blood cells agglutinates c^c-positive chimpanzee cells (i.e., of Rc or of rc type), but does not react with C^c-positive red cells (i.e., type RC or rC). Such absorption removed the anti-R^c antibodies while leaving behind the anti-c^c.

The c^c presents an antithetic reaction in relation to C^c (as do C and c in man)—i.e., when c^c is absent in an individual, C^c is sure to be found and vice versa. Some individuals possess the two factors, which implies the action of a pair of codominant alleles. In contrast, there is no antithetic reaction between E^c and F^c but the analysis of their distribution in the populations of chimpanzees studied and their mode of transmission in the families demonstrate that all the factors are genetically linked [202]. In particular, as we shall see later, the E^c and F^c factors appear only if C^c is also present on the red blood cells (whereas C^c can appear without them).

Genetic Model of the R-C-E-F System of Chimpanzee

To explain the transmission of the C-c-E-F phenotypes, A.S. Wiener et al. postulated [275] (at a time when the R^c had not yet been discovered) the existence of a polyallelic series of five codominant mutations—c, C, C^E, C^F, C^{EF}—responsible for the synthesis of either one factor (C or c) or two factors (CE or CF), or even three factors (CEF) appearing concurrently on the red cells of chimpanzees. However, a few recent findings caused us to review this model. First, the discovery of the R^c factor, mentioned earlier, which on the one hand was found to be closely associated with the factors of the C-E-F chimpanzee system and, on the other, proved to be immunologically very similar, if not identical, to the human Rh_0 (D) factor. On the phylogenic level, R^c must play the same central role for the R-C-E-F system as does the D factor of the Rhesus system in man.

The second discovery is connected to the obtaining (by alloimmunization of chimpanzees) of an antibody initially called anti-P^c [126]. This reagent was later found to be closely associated with anti-c^c: it regularly gave negative reactions when c^c was absent; but the converse was not always true, and occasionally c^c was absent when P^c was present. Their relationship seemed comparable to that of A_1 and A_2 of the A-B-O system. Thus, the anti-P^c became the anti-$^c_1$ [202]. The use of these two reagents in testing c^c-

positive red blood cells allowed the identification of two subgroups: c_1 and c_2 (see Table 7-3).

Factors C^c, E^c, and F^c are easily recognized by the corresponding antibodies, which are always highly specific as are anti-c^c and anti-c_1^c. It was found, moreover, that factors E^c and F^c are never present without C^c; yet, C^c can exist alone, with or without R. This demonstrates that C, E, and F must surely be located at the same locus.

Taking into account all the reactions that can be observed by using the different antibodies now available, two genetic models can be envisaged. The first, proposed by Socha and Moor-Jankowski [202], involves the existence of nine alleles, each one ensuring the simultaneous synthesis of one to four factors (the list of these nine alleles and the factors whose synthesis each one ensures are given in Table 7-4).

The R and r alleles ensure the synthesis of c_1 or c_2, according to the superscript they carry (R^1, R^2, r^1, r^2). R is, in addition, responsible for the synthesis of the R^c factor, which, as we explained earlier, possesses an antigenic fraction in common with the human Rh_0 (D) factor.

The alleles r^C, r^{CE}, and r^{CEF}, although theoretically possible, were not indispensable for the explanation of heredity of the R-C-E-F phenotypes thus far observed and were not included in the original model of the R-C-E-F system [202].

TABLE 7-3. Subdivision of the type c^c by means of anti-c^c and anti-c_1^c reagents

Reactions with reagent:		Phenotype
Anti-c_1^c	Anti-c^c	
+	+	c_1^c
−	+	c_2^c

TABLE 7-4 Postulated alleles of the R-C-E-F system and their products

Alleles	Factors synthesized
r^1	c_1^c
r^2	c_2^c
r^{CF}	$C^c + F^c$
R^1	$R^c + c_1^c$
R^2	$R^c + c_2^c$
R^C	$R^c + C^c$
R^{CF}	$R^c + C^c + F^c$
R^{CE}	$R^c + C^c + E^c$
R^{CEF}	$R^c + C^c + E^c + F^c$

The nine alleles initially postulated could combine into $(n + 1)n/2 = 45$ (where n is the number of alleles) possible genotypes, which theoretically could give 19 phenotypes. Of these, all but two were actually observed in a sample of 203 unrelated adult chimpanzees *P. troglodytes* (Table 7-5).

The observed distribution of the R-C-E-F blood types gave a satisfactory fit with expected frequencies, based on the nine-allele model of inheritance. The gene frequencies given at the bottom of Table 7-5 were calculated using the formulas published in 1980 [202].

As the number of chimpanzees tested has more than doubled in the last 3 years, not only the two missing, but expected, phenotypes (RC and $RCEFc_2$) were encountered, but also some additional types not originally postulated. Among those were type rCc_1 and a series of irregular forms, some of which proved to be inheritable. The list of the R-C-E-F blood types actually observed at the time of this writing is given in Table 11-13 (page 230). The newly encountered types necessitate introduction of at least one additional, regular allele, namely r^C—not to mention the irregular mutations that seem to be not infrequent among common chimpanzees.[2]

An alternative model, which is perhaps as convenient to use, assumes the existence of two sets of linked alleles. Since C, C^E, C^F, C^{EF}, c^1, and c^2 behave antithetically, they probably belong to the same allelic series. The pair R/r, on the other hand, could occupy an adjacent locus. This would result in the following arrangement of the chromosome segment carrying genes of the R-C-E-F chimpanzee blood group system:

first locus: R, R^u, r
second locus: $C, C^E, C^F, C^{EF}, c^1, c^2$

If the rare mutation R^u is disregarded, there would be 12 possible "chromosomes," of which rC^E and rC^{EF} have not been encountered. As mentioned earlier, the relatively small number of animals tested so far, and the even smaller number of chimpanzee families investigated (no more than 40 to date), account easily for the absence of some rare blood types among the animals that were available for testing.

If, as we believe, this second multilocus model is correct, it is possible that the Rhesus blood group system of chimpanzee, not unlike the human Rh system, is formed of three segments:

[2]This shows clearly how the small number of animals available for testing impairs the study of inheritance of blood groups in populations of nonhuman primates. If, until now, we had typed only fewer than 400 specimens of human blood, we would have probably missed the chromosomes Cde, cdE, and CDE, not to speak of the extremely rare CdE chromosome of the Rh blood group system.

TABLE 7-5. Serology and genetics of the chimpanzee R-C-E-F blood group system

Designation	Reaction with serum of specificity anti- R^c C^c E^c F^c c^c c_1^c						Number Observed	Expected[a]	Possible genotype(s)
rc_1	−	−	−	−	+	+	53	50.17	r^1/r^1, r^1/r^2
rc_2	−	−	−	−	+	−	3	2.84	r^2/r^2
rCF	−	+	−	+	−	−	2	2.15	r^{CF}/r^{CF}
$rCFc_1$	−	+	−	+	+	+	18	16.52	r^{CF}/r^1
rCF_2	−	+	−	+	+	−	5	4.97	r^{CF}/r^2
Rc_1	+	−	−	−	+	+	37	35.02	R^1/R^1, R^1/R^2, R^1/r^1, R^1/r^2, R^2/r^1
Rc_2	+	−	−	−	+	−	16	15.15	R^2/R^2, R^2/r^2
RC	+	+	−	−	−	−	0	0.02	R^C/R^C
RCc_1	+	+	−	−	+	+	2	1.64	R^C/R^1, R^C/r^1
RCc_2	+	+	−	−	+	−	2	1.18	R^C/R^2, R^C/r^2
RCE	+	+	+	−	−	−	2	0.80	R^{CE}/R^{CE}, R^{CE}/R^C
$RCEc_1$	+	+	+	−	+	+	6	9.05	R^{CE}/R^1, R^{CE}/r^1
$RCEc_2$	+	+	+	−	+	−	4	6.94	R^{CE}/R^2, R^{CE}/r^2
RCF	+	+	−	+	−	−	11	8.13	R^{CF}/R^{CF}, R^{CF}/R^C, R^{CF}/r^{CF}, R^C/r^{CF}
$RCFc_1$	+	+	−	+	+	+	19	19.87	R^{CF}/R^1, R^{CF}/r^1, R^1/r^{CF}
$RCFc_2$	+	+	−	+	+	−	16	13.97	R^{CF}/R^2, R^{CF}/r^2, R^2/r^{CF}
RCEF	+	+	+	+	−	−	6	5.60	R^{CEF}/R^{CEF}, R^{CEF}/R^{CE}, R^{CEF}/R^{CF}, R^{CEF}/R^C, R^{CEF}/r^{CF}, R^{CE}/R^{CF}, R^{CE}/r^{CF}
$RCEFc_1$	+	+	+	+	+	+	1	1.32	R^{CEF}/R^1, R^{CEF}/r^1
$RCEFc_2$	+	+	+	+	+	−	0	0.95	R^{CEF}/R^2, R^{CEF}/r^2

$$\chi^2_{(10)}=7.6023;\ 0.70>P>0.50.$$

[a]Based on estimated gene frequencies:

$r^1=0.39272$	$r^2=0.11828$	$r^{CF}=0.10357$
$R^1=0.02254$	$R^2=0.17983$	$R^C=0.00974$
$R^{CE}=0.05369$	$R^{CF}=0.11225$	$R^{CEF}=0.00783$

The *first*, which carries R or r, corresponds to the human D/D^u/d segment, with which it has many common sequences.

The *second* carries the gene responsible for the synthesis of the hr' (c) factor, present both in man and chimpanzee. In the latter, however, unlike the situation in man, this locus is not polymorphic.

The *third*, which carries the allelic series C, C^E, C^F, C^{EF}, c^1, c^2, might correspond to the E/e locus of man. But here the two chromosome segments would each have evolved differently in man and in anthropoid apes.

The comparison between the human chromosome segments and the corresponding chromosome segment in anthropoid apes can thus be shown as follows:

Man	*Chimpanzee*
D, or D^u, or d	R, or R^u or r
rh′, or hr′ (C/c)	hr′ (c)
rh″, or hr″ (E/e)	C, or C^E, or C^F, or C^{EF}, or c^1, or c^2

As shown by population analysis, whatever the genetic model chosen, there is good agreement between the observed and expected frequencies, even in this limited-size sample of chimpanzees studied.

The ten pigmy chimpanzees *(Pan paniscus)* so far tested with the battery of R-C-E-F typing reagents were all found to be type RCE. As the data in Table 7-5 indicate, the type RCE is relatively rare among common chimpanzees.

It may be of interest that, unlike chimpanzees of the common type, which are highly polymorphic for all known chimpanzee blood groups, the dwarf chimpanzees are surprisingly uniform with respect to their blood groups. This observation applies not only to animals born in captivity but also to wild animals [195a]. For more details the reader is referred to the section on seroprimatology (pp 212–219).

The R-C-E-F System in Other Nonhuman Primate Species

Gorillas. All gorillas, both lowland *(Gorilla gorilla)* and mountain *(Gorilla beringei)*, are R^c-positive [282]. This monomorphic factor of gorilla red cells appears either alone, which could be considered the equivalent of the "-D-" type in man, or accompanied by the C or CF factors. The c^c allele, either in its c^1 or c^2 form, has never been observed in these apes, nor has the E^c specificity. Therefore, if the observations on relatively small numbers of animals are valid, the existence of three mutations, R, R^C, and R^{CF}, at the R-C-E-F locus of gorillas may be assumed. If instead of the polyallelic unilocus model of the inheritance of the R-C-E-F groups one considers an alternative model consisting of a series of closely linked genes, two loci can be distinguished:

first locus: R or r

second locus: C or C^F or c

where r and c are silent alleles.

Obviously, since the number of gorillas thus far investigated does not exceed 40, there is a good chance that still further R-C-E-F alleles will be identified, or that a phenotypic expression of genes so far considered silent alleles will become known.

Leaving aside differences in the numbers of gorillas and chimpanzees tested, there are striking peculiarities in the distribution of the R-C-E-F blood groups in either species: 1) the solitary type R has never been observed among chimpanzees, while two-thirds of gorillas tested are of this type (and the rest are types RC or RCF); 2) type RC is extremely rare among common chimpanzees (in fact the first two *Pan troglodytes* to have RC type red cells have been only recently encountered) but quite frequent among gorillas.

Moreover, the R-C-E-F antigens on gorilla red cells are always weaker than those of the chimpanzees [200]. In fact, absorptions of anti-R^c, anti-C^c, and anti-F^c reagents with positively reacting chimpanzee red cells rendered the sera inactive against chimpanzee as well as gorilla erythrocytes. The reverse, however, was not true: the same sera absorbed with gorilla red cells of type RCF were still able to agglutinate the chimpanzee red cells of the corresponding types. These differences seem secondary, and we believe that the animals of both species of African apes carry the same chromosome sequence, which also corresponds to the chromosome segment responsible for the antigenic red cell properties of the Rh system of man.

Gibbons. The red blood cells of 12 gibbons gave uniformly positive reactions with anti-c^c antibodies. In tests with anti-R^c, 10 animals were found to be positive and 2 negative. The tests with the other R-C-E-F reagents (anti-C^c, anti-c_1, anti-E, and anti-F^c) were all negative.

Orangutans. We have said that the results with blood specimens of orangutans *(Pongo pygmaeus)* were difficult to interpret because of interference from potent anti-orang heteroagglutinins normally present in chimpanzee sera. These are impossible to absorb without causing the titer of type-specific antibodies to drop. No readable reactions were obtained with the red cells of six orangutans previously tested.

Recently, however, large quantities of blood became available from several members of an orangutan family. Multiple absorption, fractionation, and elution experiments, carried out in parallel with orangutan and chimpanzee red cells, led to the conclusion that at least two R-C-E-F types occur in orangutans, namely RCc_1 and rc_1 (for further details, see the discussion of erythroblastosis fetalis, pp. 179–185).

Old World monkeys. None of the factors of the R-C-E-F system could be established on the red cells of primates that are situated lower on the phylogenic ladder and that separated earlier from the common trunk. All tests carried out with blood specimens of macaques, baboons, vervet monkeys, and *Cercopithecus* remained negative.

THE RHESUS SYSTEM IN EVOLUTION

The preceding data suggest that the Rhesus system, like the M-N system, is formed of several sequences that did not appear simultaneously in evolution but in successive stages.

The oldest sequences (which are found in virtually identical forms in anthropoid apes and in man) involve factors D/R^c and hr' (c). They must have been already present in the common ancestor and have scarcely changed since.

Another sequence, which could be due to a more recent duplication of a chromosome segment, had probably evolved in two different directions, leading to the $E/e/E^W/e^i$ series in man and the $C^c/c^c/E^c/F^c$ series in anthropoid apes. A comparable phenomenon was described, as we remember, in the M-N system. This demonstrates the importance of the redundancies of chromosome sequences in the diversifying evolution of species. This evolutionary model can be represented by the following diagram:

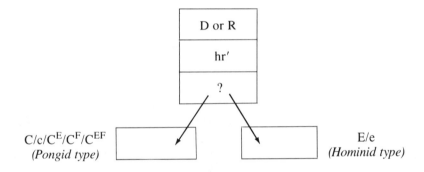

$C/c/C^E/C^F/C^{EF}$ D or R   E/e
(Pongid type)   hr'   *(Hominid type)*

Blood Groups of Primates, pages 91–98
© 1983 Alan R. Liss, Inc., 150 Fifth Avenue, New York, NY 10011

Blood Systems Specific to Cercopithecoidea (Old World Monkeys)

The Graded D^{rh} Blood Group System of Macaques 91
The Graded B^p System of Baboons 94
Relationship of the Graded Blood Group Systems of Old World
   Monkeys to Those of Anthropoid Apes and Man. 96

All species of primates so far studied, which for the most part involve the Old World monkeys (Cercopithecidae), have revealed the presence of blood factors particular to each species [124,127] or present in several species closely related phylogenetically (para-antigens) [173]. In the majority of cases, these antigens were revealed either by alloimmunization or by heterospecific antibodies. Unfortunately, given the present stage of our knowledge, it is not possible to know if they are attached to the blood systems that we have described in the previous chapters and that were first discovered in man and then reencountered, in a modified state, in some anthropoid apes, or if they are autonomous systems that have no counterparts in man or apes. Here we will limit ourselves to reviewing the now well-identified systems, without prejudging in any way the links they may have with those already known in anthropoid apes and man.

THE GRADED D^{rh} BLOOD GROUP SYSTEM OF MACAQUES

Macaca mulatta was the subject of alloimmunization procedures that revealed at least 25 different specificities, some of which belonged to one genetic system [211,287]. The most remarkable and probably the most important is the D^{rh} system, which was first found in the rhesus monkey (*Macaca mulatta*) but is also present in a certain number of related species of macaques [125].

This system presents a series of factors—D_1^{rh}, D_2^{rh}, D_3^{rh}, D_4^{rh}—that correspond to as many interlinked grades of one fundamental antigen. All grades of the system that are detected in rhesus monkeys are also present in pig-tailed macaques (*Macaca nemestrina*) [204] and in crab-eating macaques (*M. fascicularis*) (unpublished). The D^{rh} system appears in somewhat truncated form in stump-tailed macaques (*M. arctoides*) [124], Barbary macaques (*M. sylvanus*) [198], and hamadryas baboons [125]. Tests for the specificities of the graded D^{rh} system were negative with the red cells of all bonnet macaques (*M. radiata*) [205] and Japanese macaques (*M. fuscata*) (unpublished) investigated to date.

For the sake of consistency, we now adopt the following nomenclature, which differs a little from that initially used and has been established as new factors have been discovered [207].

D_1^{rh} group carries the D_1 specificity.
D_2^{rh} group carries the $D_1 + D_2$ specificities.
D_3^{rh} group carries the $D_1 + D_2 + D_3$ specificities.
D_4^{rh} group carries the $D_1 + D_2 + D_3 + D_4$ specificities
  and corresponds to the "standard" D type, the first to
  be discovered.

The four groups of the D^{rh} system so far known are defined by means of a battery of four anti-D^{rh} reagents; the list of the anti-D^{rh} sera and their description are given in Table 8-1. Table 8-2 explains the serology of the graded D^{rh} system and shows the distribution of the phenotypes among some species of the Old World monkeys.

The comparison between the reactions observed in *M. mulatta, M. nemestrina,* and *M. fascicularis,* using four anti-D^{rh} antisera, gives the idea of a "step-like" evolution, which was discussed above. In *M. arctoides* and in *Papio hamadryas,* there exist special mutations, designated D_3' and $D_3'',$ that are characterized by the absence of antigenic sequences found in the preceding groups (D_2 is missing in the D_3' type, while D_1 is missing in the D_3'' type). Because of the small number of animals studied, we cannot be certain that these mutations are absent in other species.

TABLE 8-1. Anti-D^{rh} typing reagents

Designation of reagent	Specificity of antibodies
Anti-D standard (anti-D_1)	Anti-D_4, anti-D_3, anti-D_2, anti-D_1
Anti-D_2	Anti-D_4, anti-D_3, anti-D_2
Anti-D_3	Anti-D_4, anti-D_3
Anti-D_4	Anti-D_4

TABLE 8-2. Serological definition of blood groups of the D^{rh} graded system and their distribution among various species of Old World monkeys

Blood group designation	Specificities present on the red cells	Reactions of red cells with reagent:				Frequencies in Old World monkey species:						
		Anti-D_1	Anti-D_2	Anti-D_3	Anti-D_4	M. mulatta	M. nemestrina	M. fascicularis	M. arctoides	M. radiata	M. sylvanus	Papio hamadryas
D_4 (D standard)	D_1,D_2,D_3,D_4	+	+	+	+	34.2	34.9	8.7	0.0	0.0	0.0	0.0
D_3	D_1,D_2,D_3	+	+	+	−	6.3	2.2	4.0	0.0	0.0	78.1	0.0
D_2	D_1,D_2	+	+	−	−	23.0	45.0	21.4	3.3	0.0	21.9	0.0
D_1	D_1	+	−	−	−	1.6	5.6	23.0	0.0	0.0	0.0	41.0
d		−	−	−	−	34.9	12.3	42.9	66.7	100.0	0.0	0.0
Irregular types:												
$D_3^{.}$		+	−	+	−	0.0	0.0	0.0	0.0	0.0	0.0	59.0
D_3		−	+	+	−	0.0	0.0	0.0	30.0	0.0	0.0	0.0
Number of animals tested						126	89	126	30	52	32	52

From the genetic point of view, each D^{rh} blood type corresponds to an autonomous mutation; when combined, these form a series of "step-like" alleles. Each new factor implies the presence of preceding factors, to which new sequences are added.

The direction of dominance goes from the most complex to the most simple factors; finally, a recessive d^{rh} mutation exists that is considered a silent allele because it produces no detectable factor:

$$D_4^{rh} > D_3^{rh} > D_2^{rh} > D_1^{rh} > d^{rh}$$

The frequencies observed in large samples of unrelated rhesus monkeys and pig-tailed macaques agreed well with theoretically expected frequencies based on that model of inheritance by a series of multiple graded alleles. The explanation of that model and its analysis, carried out on a population of pig-tailed macaques (*Macaca nemestrina*), are shown in Table 8-3.

THE GRADED B^P SYSTEM OF BABOONS

The B^P system, in many respects comparable to the D^{rh} system of macaques just described, has just been found in hamadryas baboons (*Papio hamadryas*) [203], but its existence was also confirmed in other species of baboons, namely olive baboons (*P. anubis*) and yellow baboons (*P. cynocephalus*). It has not yet been tested in other species of monkeys.

This system also displays the presence of several graded types:

B_4^P (or B^P standard) contains four factors: B_1, B_2, B_3 and B_4.
B_3 contains three factors: B_1^P, B_2^P and B_3^P.
B_2 contains two factors: B_1^P and B_2^P.
B_1 contains only the factor B_1^P.

Finally, a B^P-negative type (b^P) occurs, probably as a product of a recessive mutation.

It is assumed that the graded B^P system is the result (as is the D^{rh} system of macaques) of several progressive "step-like" mutations, in which each new mutation contains more antigenic sites and dominates over the previous mutation, which contains fewer antigenic sites. In this way, the following order of dominance is established:

$$B_4^P > B_3^P > B_2^P > B_1^P > b^P$$

The graded types of the B^P systems are defined by a set of anti-B^P re-agents—anti-B_1^P, anti-B_2^P, anti-B_3^P, and anti-B_4^P—each containing antibodies for one or more specificities characterizing different grades of the B^P antigen. All B^P reagents were obtained by intentional alloimmunization, but several "normal" sera of apparently nonimmunized baboons were discovered containing spontaneous antibodies of various anti-B^P specificities. Some of these "natural agglutinins" were probably the result of transplacental immunization in the course of incompatible pregnancies (see page 166), but the origin of some others, particularly those found in the sera of wild male baboons (unpublished observation), has remained a mystery.

Although the genetic model of the B^P system is very similar to that of the D^{rh} graded system, the two systems are immunologically different since they originate in taxonomically distinct groups. Nevertheless, there are indications that the B^P antigen of the baboon red cell is somewhat related to the D^{rh} of macaques: it has been established that the anti-D_3^{rh} antibody recognized the B_3^P factor on the baboon red cell. The reverse, however, is not true.

It appears, therefore, that there is a certain phylogenic connection between the D^{rh} macaque system and the B^P blood groups of baboons, in the form of a common antigen zone situated at the level of D_3 subgroup of the first system and at B_3 for the second.

The anti-B_3^P antiserum, as noted above, is inactive against the red cells of macaques. It is noteworthy, however, that anti-B_4^P was found to detect polymorphic specificity (completely independent from the D^{rh} system) on the red cells of *Macaca mulatta* and *M. nemestrina*, and also in the blood of geladas (*Theropithecus gelada*). This further strengthens the idea of an immunological kinship among species of the Old World monkeys.

The relationship between the two systems can be visualized in Figures 8-1 and 8-2.

TABLE 8-3. Genetics of the D^{rh} system as observed in a sample of unrelated pig-tailed macaques (*Macaca nemestrina*)

Phenotype	Possible genotypes	Number	
		Observed	Expected
D_1	D^1D^1, D^1d	1	0.98
D_2	D^2D^2, D^2D^1, D^2d	18	17.99
D_3	D^3D^3, D^3D^2, D^3D^1, D^3d	2	2.01
D_4	D^4D^4, D^4D^3, D^4D^2, D^4D^1, D^4d	12	12.01
d	dd	8	7.99

Gene frequencies: $D^4 = 0.1592$; $D^3 = 0.0296$; $D^2 = 0.3432$; $D^1 = 0.0264$; $d = 0.4416$.

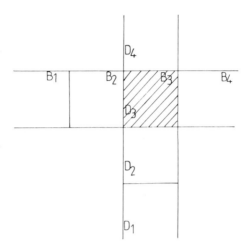

Fig 8-1. Two immunological stocks, one corresponding to the baboon phylum and the other to the rhesus phylum, may have emerged from a common ancestor (represented by the lined area) that had antigenic structures, shared by B_3 of baboon and D_3 of macaques.

1. Figure 8-1 shows how, from a single ancestral antigen (common parts corresponding to B_3^p/D_3^{rh}), two "immunological stocks" were differentiated—one corresponding to the B_1, B_2, B_3, B_4 of baboons and the other to the D_1, D_2, D_3, D_4 of macaques.

2. Figure 8-2 shows the relationships that link the different immunological types within each lineage: D_4 has all the patterns of D_1, D_2, and D_3 in addition to its own; D_3 has all the patterns of D_1 and D_2 in addition to its own, etc. This represents an evolution "in steps that overlapped each other."

The B^p system of baboons displays the same evolutionary type.

RELATIONSHIP OF THE GRADED BLOOD GROUP SYSTEMS OF OLD WORLD MONKEYS TO THOSE OF ANTHROPOID APES AND MAN

The R-C-E-F system of anthropoid apes corresponds, as we have seen, to the human Rhesus system. It constitutes its equivalent, but it is differently oriented. The meeting point between the two (or, if one prefers, their common starting point) corresponds to common antigenic patterns that are part of the R^c factor of apes and of the human Rh_0 (D) factor. In a similar way, the V-A-B-D chimpanzee system is the equivalent of the human M-N-S-s system. Here again, the N^V factor (identical to V^c) represents the "immunological crossroads" from which the antigens particular to each lineage were differentiated.

But can we go further back in phylogeny? As an example, let us look at the Rhesus system. Given its immunological and genetic complexity, it is probable that the genetic information that led to the R-C-E-F of the chimpanzee and its homologue, the human C-D-E system, did not appear all at once in the first group of monkeys that emerged, probably at the beginning of the Miocene era, from which hominids and anthropoids originated.

Before the current level of development of the two blood group systems of Hominoidea was achieved, some equally complex structures must have existed that crossed successive stages; and the oldest primate species still living must have retained some similarities to these structures. It is therefore possible that the Rhesus system had emerged already, at least in a rudimentary form, in the Old World monkeys (macaques or baboons) and that antigenic patterns exist that are common to different lineages. Thus far, no immunological relationship has been demonstrated between the D^{rh} and B^P systems on one hand and the R-C-E-F and human Rhesus systems on the other. Perhaps no relationship exists. Maybe the common factor that bridges the immunological systems of monkeys and those of anthropoid apes and man (similarly to

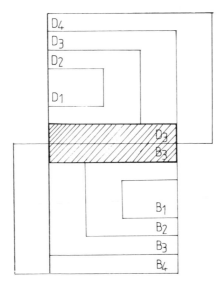

Fig 8-2. Diagram showing how the "step-like" mutations that invoked a common ancestral factor (lined area) led to the formation of the graded subtypes, either by acquisition of new factors (toward B_4 and D_4 types) or by loss of certain patterns (B_1 and D_1 types).

the way the specificity R spans the human Rh and the chimpanzee R-C-E-F system) still remains to be discovered. However, the existence of such relationships cannot be doubted; we must not forget that it was by immunizing rabbits with red blood cells of rhesus monkeys that the antisera were produced that led to the discovery of one of the most important antigenic systems of man, the Rhesus system.

But the antigen that Landsteiner and Wiener described was not absolutely comparable to the blood factor recognized by human anti-Rh, and therefore the anti-monkey serum was assigned by Levine the symbol anti-LW. In fact, what is striking when the immunization experiments by Landsteiner and Wiener are reexamined is the complexity of the antibodies obtained: a mixture of several antibodies, whose relative proportions, moreover, can change from one experiment to another. It seems that the red blood cells of the macaques contain a complete mosaic of antigens capable of provoking the appearance in the rabbit of a series of antibodies, at least six of which have been exactly identified: anti-LW; anti-M; anti-*Macaca mulatta*; anti-*Papio*; antibodies active simultaneously both against macaque and *Papio* cells and against human cord cells; and antibodies cross-reacting with macaque and *Papio* and adult human red cells, etc. Is there, in this collection, an antibody able specifically to recognize the human D factor, whose existence was still affirmed by A.S. Wiener [262] but has been contested by many immunologists? In the present state of our knowledge of the blood groups of nonhuman primates we cannot hazard a final opinion.

Blood Groups of Primates, pages 99–161
© 1983 Alan R. Liss, Inc., 150 Fifth Avenue, New York, NY 10011

Methodology of Blood Grouping in Nonhuman Primates

Development of Testing Techniques 99
Collection and Shipment of Specimens for Blood Grouping 106
Tests With Reagents Originally Prepared for Typing Human Blood 109
    Tests for A-B-O Blood Groups 109
    M-N Typing 120
    Rh-Hr Typing 120
    Lewis Typing 128
    Testing for I and i Blood Factors 129
    Tests for Homologues of Other Human Blood Groups 132
Tests With Reagents Produced by Immunization of Primate Animals 132
    Immunization 133
    Testing of Antisera 138
    Production of Antiglobulin Sera 139
    Immunization Protocol 141
    Plasmapheresis 143
    Standardization of Typing Reagents 143
    Typing Reagents 153
Monoclonal Antibodies 158

DEVELOPMENT OF TESTING TECHNIQUES

The earliest investigations of blood groups of nonhuman primates were concerned with detection of A and B agglutinogens on the red blood cells of the animals, as has been discussed in earlier chapters.

For the A-B-O grouping of human blood, the red cells are tested with anti-A and anti-B sera, and these results are corroborated by reverse grouping of the individual's serum against known cells of groups O, A_1, A_2, and B. However, when used in tests on the red cells of nonhuman primates, the anti-A and anti-B sera, prepared in the usual manner from sera of immunized or nonimmunized human individuals, have been found to contain heteroagglutinins reacting with red cells of *all* apes and monkeys, which obscures the group-specific reactions. The reverse grouping tests on the nonhuman primate sera, on the other hand, do not present this difficulty, since when species-specific heteroagglutinins for human red cells are present, they are in most cases weak, and their activity is generally readily eliminated by dilution.

If human anti-A and anti-B are to be used for grouping red cells of nonhuman primates, steps must be taken to avoid the interference of the species-specific heteroagglutinins. Otherwise, reactions caused by such heteroagglutinins may be mistaken for group-specific reactions, and errors in grouping result. Some of the early workers, for example, used human anti-A and anti-B directly for typing the blood of chimpanzees and found some of the animals to be group AB, a blood group that does not occur in this species. In the earliest significant investigations on the blood groups of apes, Landsteiner and Miller succeeded in overcoming the interference of heteroagglutinins by the use of eluates [93]. They treated human A cells with anti-A sera and then eluted the anti-A agglutinins from these cells. The anti-A eluates, as well as anti-B eluates prepared by elution of anti-B from B cells, were then used for grouping blood of various species of anthropoid apes.

A second method employed by Landsteiner and Miller was immunization of rabbits with human group A or group B cells and preparation of specific anti-A and anti-B testing reagents from immune sera by absorption with human O cells. Following Landsteiner's rule of "immunological perspective," such absorptions removed heteroagglutinins directed not only against humans, but also against closely related apes. Unfortunately, to eliminate the very strong heteroagglutinins from rabbit immune sera, multiple absorptions were usually necessary that resulted in considerable weakening of the type-specific antibodies. It was Wiener who first demonstrated that the cumbersome eluates of immune rabbit sera could be successfully replaced by potent anti-A and anti-B reagents of human origin that had been prepared from sera of volunteers who had been immunized by intramuscular injections of commercial A and B blood group substances [250]. Because of their high titers, these sera could be used in dilutions at which the heteroagglutinins for ape red cells were not active. Another refinement proposed by Wiener et al. [264] was the use of potent immune human anti-A and anti-B sera that were rendered specific by absorption with chimpanzee group O red cells. This maneuver produced the reagents that could be used in full strength and were group-specific not only for chimpanzees, but as a rule also for orangutans and gibbons.

Further refinement of the A-B-O blood grouping techniques of apes was brought about with the introduction of lectins—as, for example, anti-A lectin of lima beans (*Phaseolus vulgaris*), anti-A$_1$ lectin (*Dolichos biflorus*), and anti-H (*Ulex europaeus*). The advantage of the use of lectins is that they do not contain heteroagglutinins and can therefore be used without prior absorptions. The same is true of the very potent anti-A snail (*Helix pomatia*) extracts or eel anti-H sera. The applications of anti-A$_1$ reagents allowed the definition of subgroups of A and AB in chimpanzees, gibbons, and orangutans, while the use of anti-H reagents contributed important information on the distribution of the H specificity in various primate species. This, in turn, combined with observations in various human races [283], led to the proposal of a new hypothesis concerning the relationship of H to the A-B group substances [285].

In their pioneer experiments, Landsteiner and Miller did not succeed in elucidating the A-B-O blood groups in gorillas or in monkeys because the red cells of those primates failed to react when crossmatched with the sera of animals of the same species or when tested with the then available reagents of anti-A or anti-B specificities [94]. Failure of monkey red cells to react with anti-A and anti-B sera was confirmed later by Buchbinder, who investigated large numbers of rhesus monkey blood specimens [16]. Nevertheless, there were indications that some kind of A-B-O differentiation exists also in monkeys—as suggested, for instance, by early observations of von Dungern and Hirszfeld [33]: after absorptions with human group O red cells, the sera of rhesus monkeys distinctly agglutinated human erythrocytes of group A but not those of group B or O. Presence in rhesus sera of hemagglutinins indistinguishable from human anti-A isoagglutinins was also confirmed by Buchbinder [19].

It was only 17 years later that Wiener et al. demonstrated that despite the absence of hemagglutination, the A-B-O type of monkeys can still be established from the presence of the A-B-H group substances in the body fluids and secretions, notably in saliva [247]. In fact, among hundreds of apes and monkeys tested so far, all proved to be A-B-H secretors, except for a single orangutan. In most cases, the results of tests on saliva of monkeys can be confirmed by testing their serum for the presence of anti-A and/or anti-B agglutinins: Landsteiner's rule[1] holds also in monkeys, except that the reciprocal relationship is between agglutinins in the serum and blood group substances in saliva and other secretions, and not between serum agglutinins and red cell agglutinogens, as is the case in man, chimpanzee, orangutan, and gibbon.

---

[1]When an isoagglutinogen is lacking in the blood cells, the corresponding isoagglutinin is present in the serum of the same blood.

The saliva inhibition technique was also successfully used by Wiener et al. [253] for establishing the Lewis types of saliva from many species of apes and monkeys. In contrast, tests for Lewis types of red cells of nonhuman primates have not been possible because of lack of potent antisera free of nonspecific heteroagglutinins. The presence of such heteroagglutinins in anti-Lewis sera of human origin does not interfere, on the other hand, with inhibition tests in which human red cells are used as indicator cells.

Rabbit immune antihuman sera were employed by Landsteiner and Wiener in direct hemagglutination tests to detect M and N agglutinogens on the red cells of apes and monkeys [97]. Though not devoid of strong heteroagglutinins, the anti-M and, to a lesser degree, the anti-N rabbit sera were potent enough to be used in absorption and titration-dilution experiments that proved the true type specificity of their reactions with primate red cells [240]. It was in those early experiments that Wiener demonstrated the complex nature of M-N antigens and the need for the use of batteries of anti-M and anti-N reagents for parallel testing with the blood of nonhuman primates. Reagents prepared from different rabbit antisera, though of identical specificity in tests on human red cells, differed in their reactions with red cells of apes and monkeys. Therefore, when tests are carried out on nonhuman primate red cells using a single reagent, the results obtained can be misleading. Some of the earlier investigators carried out their tests on red cells of rhesus monkeys with a single anti-M reagent, and therefore conflicting results were obtained [24,89]. A few workers concluded that red cells of rhesus monkeys were M-positive; others insisted that they were M-negative. The results of tests on red cells of nonhuman primates with anti-N rabbit sera were analogous to those obtained with anti-M reagents, though they were, by and large, less satisfactory because the available anti-N reagents were of lower titer and had a greater tendency to produce nonspecific reactions. The discovery by Ottensooser and Silberschmidt [155] that extracts of seeds of *Vicia graminea* have anti-N specificity has proved most helpful in this connection, and this anti-N^V lectin, free of the interfering heteroagglutinins, has been used routinely in studies of M-N groups of nonhuman primates. Specificity of reactions of anti-N lectin with chimpanzee red cells was confirmed in a series of experiments by Levine et al. [107]. Another source of anti-N lectin applicable for tests on nonhuman primate blood was found in extracts of the leaves of the Korean *Vicia unijuga* [123].

The importance of the choice of proper methodology in testing blood of nonhuman primates is well illustrated by the early attempts to discover Rh agglutinogen on the red cells of chimpanzees. The first such investigation was that of Wiener and Wade [291]. They reported all of a series of ten chimpanzees to be Rh-negative, since the red cells were not agglutinated by anti-Rh_0 but reacted with anti-hr' serum. That observation was confirmed soon afterwards by Mourant and Race [148]. Not until 1952 did Wiener find, in tests on a single chimpanzee, that the negative reactions obtained in his original study must have been due to the use of "saline agglutinating" anti-Rh_0 serum, which did not clump chimpanzee red cells by the saline agglutinating method, the only technique available at the time of Wiener and Wade's original study. However, with the newly introduced technique of enzyme treatment it was discovered [241] that the ficinated chimpanzee red cells were agglutinated by anti-Rh_0 as well as by anti-hr' sera of human origin. Further tests are still necessary to determine whether the reagents contain nonspecific heteroagglutinins for ape blood; no simple method has been devised to remove them without weakening the type-specific antibodies. Yet, when the reagents give negative reactions in direct tests on ape blood, the specificity being tested for is obviously absent from the tested cells. If, on the other hand, positive reactions are obtained, further tests are required to establish whether these are type-specific or are due to other antibodies in the reagent. One of the methods proposed by Wiener et al. [248] is titration of the positively reacting simian red cells in parallel with positive human red cells. Since the reagents used for these experiments are of high titer, similarity of titers may be considered evidence that the reactions are indeed type-specific. A more definitive proof is absorption of the reagent, anti-Rh_0 for example, with human Rh_0-positive red cells. If this removes reactions for the ape red cells also, the reaction of the unabsorbed serum is considered type-specific.

Potent Rh reagents for typing ape blood are prepared from sera of isoimmunized human patients, or from male Rh-negative volunteer donors immunized with Rh-positive blood. A specificity identical or closely related to Rh_0 can also be identified on the red cells of various species of anthropoid apes by means of reagents prepared from the sera of isoimmunized chimpanzees [199].

Blood factors of primate red cells that could also be investigated with reagents originally prepared for typing human blood were factors I and i. By means of extremely potent antisera obtained from patients with autohemolytic anemia and freed from species-specific heteroagglutinins by simple dilution, red cells of a number of nonhuman species—apes and monkeys—could be tested for the presence of factor I [132] and then of factor i [272].

The reagents for testing most other human blood group systems are generally of low titer and avidity. Therefore, the results are poorly reproducible even in man and are particularly unsuitable for testing simian red cells because of interference of nonspecific heteroagglutinins. The reports in the literature claiming to have demonstrated specificities such as Kell, Kidd, Duffy, etc., on the red cells of nonhuman primates are mostly based on simple direct tests or tests with eluates, and they significantly make no mention of confirmatory tests by titration and absorptions, or the use of the blind technique; they are thererfore unreliable.

Introduction of reagents produced by immunization of primate animals or prepared by immunization of other animals with the red cells of nonhuman primates not only enlarged the scope of blood grouping tests available for testing apes and monkeys, but also opened a new and large territory for theoretical and practical applications of information obtained with these new serological tools. Three approaches have been used to produce this type reagents: 1) isoimmunization, i.e., injection of primates with red cells from animals of the same species; 2) cross-immunization, defined by Moor-Jankowski and Wiener [130] as immunization between closely related species, e.g., immunization of chimpanzees with red cells of man or gibbon; and 3) heteroimmunization, i.e., injection of rabbits or other laboratory animals with primate red cells.

Because of antigenic differences between animals widely separated taxonomically, production of heteroimmune antibodies seemed by far the easiest. The resulting antiserum, however, contained a mixture of antibodies of many various specificities, including also potent species-specific antibodies, which are very often hard to separate. Attempts to fractionate such antisera often result in very weak final products that give hardly reproducible results. This was the case of antisera produced by Owen and Anderson by immunization of chickens with red cells of rhesus monkeys [156]. A more reliable test system was obtained by these same authors with rabbit antirhesus sera of five different specificities, identified by absorption analysis. No attempts, however, were made to ascertain that the reagents were monospecific in either a serological or a genetic sense, and the study was not continued. Production of rabbit antimonkey sera was taken up by Bogden et al. [11], and in another study by Foran [46]. Foran obtained heteroimmune sera that were reported to detect on rhesus monkey red cells at least six specificities that were different from those described earlier by Owen and Anderson. The reagents of Foran cross-reacted with red cells of Celebes apes (*Macaca maura*). This work was later pursued by LaSalle and Frisch [104], but no details were given regarding monospecificity of the antisera and genetic interrelationship among the blood group specificities defined by them. The earliest trial to produce typing reagents by immunizing a monkey with the red cells of an animal of a closely related species was that of Fischer and Klinkhart [45]. Serum obtained from a rhesus immunized with red cells of the cynomolgus monkey (*Macaca fascicularis*) detected individual differences among rhesus monkeys and also cross-reacted with red cells of cynomolgus monkeys. In another experiment, Landsteiner and Levine [92] immunized three chimpanzees with A-B-O–compatible human red cells and produced antisera that detected individual differences in chimpanzee red cells. Type-specific anti-human chimpanzee sera are now part of the batteries of reagents used for routine blood grouping tests in anthropoid apes (see Table 9-28). Similarly, cross-immune sera produced in baboons injected with the red cells of geladas are routinely used for typing baboon blood (see Table 9-27).

Understandably, the reagents of sharpest specificity can be expected to result from immunizations among animals of a single species. Several reports emphasized the difficulty of producing antisera by isoimmunization of non-human primates. It became accepted among immunologists that it was extremely difficult, if not impossible, to obtain type-specific immune sera by immunization within species or even between closely related species [7]. In more recent years, this belief was supported by the inconclusive attempts at rhesus isoimmunization by Owen and Anderson [156]. It was in the 1960s that Wiener et al. demonstrated that the prolongation of the course of immunization beyond a year or more may be required to produce adequate isoimmune primate antisera [265]. At about the same time, an immunization protocol was established that introduced the use of Freund's adjuvant to prepare the inoculum—a procedure that considerably enhanced the response of the immunized primate animals [137]. In addition, new techniques were adapted for use in tests on primate blood, namely, antiglobulin and enzyme treated red cell methods—improvements that resulted in discovery of several new so-called simian-type specificities on red cells of chimpanzees [275], gibbons [128], macaques [134], baboons [135], etc.

Iso- and cross-immunizations of primate animals initiated by Wiener and collaborators were successfully carried out by several other investigators. Among others, a team of researchers from Duke University resorted to immunization of chimpanzees with human red cells to produce cross-immune typing reagents that defined two blood factors on chimpanzee red cells [294,295]. One of these factors seemed to be somewhat related to the so-called V^c factor described by Wiener as part of the chimpanzee V-A-B-C blood group system (see page 69). Rhesus monkeys were the species of nonhuman primates most often involved in immunization attempts. Isoimmunization of rhesus monkeys as well as cross-immunization with the red cells of Celebes apes (*Macaca niger*) were used by LaSalle to obtain blood grouping reagents that detected three factors of rhesus blood [100–102]. Hirose and Balner [64], by immunizing rhesus monkeys with blood and/or skin of other animals of the same species, produced isoimmune antisera that recognized a number of red cell specificities, some of which appeared to be shared also by leukocytes (see page 205). Large-scale immunization programs were undertaken by Edwards in an attempt to produce rhesus red cell typing reagents. Saline agglutinating antisera were obtained by injecting rhesus red cells intravenously into rhesus monkeys, pigtailed macaques (*Macaca nemestrina*), stump-tailed macaques (*Macaca arctoides*), and rabbits [38–40]. The same immunization techniques, and by and large the same animals, were used by Duggleby et al. in another series of experiments that resulted in production of rhesus blood grouping reagents of various specificities [30,31]. For more detailed discussion of Edwards' and Duggleby's findings, see page 153. Production of antisera of various specificities by isoimmunization of rhesus monkeys, baboons, and mandrills was reported by Verbickij [233], but no details of experiments or discussion of specificities detected by those reagents were given. The first successful attempts at isoimmunization of New World monkeys were reported by Gengozian, who defined a number of blood factors in marmosets (*Sanguinus fuscicollis*) [52,53,55].

Further improvement of primate blood grouping techniques was brought about by introduction of low ionic strength solutions (LISS) as the red cell suspending medium [42,143], which considerably enhanced the strength and avidity of reactions with monkey sera.

Although most of the work on blood groups of nonhuman primates was carried out using serological methods developed for typing human red cells, a few trials were made to adapt, for this purpose, other serological techniques. Among others, the dextran method, originally devised for the detection of mice antibodies [60] and later modified for hemagglutination tests in dogs [171], was used by Hirose and Balner [64] for hemagglutination tests on red cells of rhesus monkeys. Dextran was also used by LaSalle [100,101] in her attempts to type isoantigens in various species of macaques. The dextran technique has proved to be unreliable in tests on human red cells and has therefore been abandoned in favor of the antiglobulin technique.

COLLECTION AND SHIPMENT OF SPECIMENS FOR BLOOD GROUPING

Collection of blood and/or saliva samples requires restraining of the animal either manually or by general anesthesia. Manual restraint is possible only in cases of very small or weak animals. More often, general anesthesia is necessary to immobilize the animal for the duration of the procedure. Nonhuman primates are usually tranquilized by intramuscular injections of anes-thetics. Two kinds of tranquilizers are most commonly used in apes and monkeys for the purpose of short anesthesia.

Sernylan (phenycyclidine hydrochloride) in 100 mg/ml concentration and 1 mg/kg dosage is usually effective for 15–20 min. Its anesthetic effects are as follows: 1) depression of the central nervous system; 2) simple reflexes (palpebral, corneal, and pupillary) not completely eliminated; 3) eyes remain open; 4) muscle tone is increased; 5) respiration and blood pressure not depressed unless the animal is overdosed.

Vetalar (ketamine hydrochloride) in 100 mg/ml concentration and 10 mg/kg dosage is usually effective for 5 min. Its anesthetic effects are as follows: 1) "dissociative" anesthesia (unconsciousness); 2) profound analge-sia; 3) protective reflexes (coughing, swallowing) are maintained; 4) a slight increase in blood pressure and decrease in respiration; 5) muscle tone is variable, depending on the individual animal; 6) eyes remain open; simple reflexes not completely eliminated.

The correct dose of anesthetics can be calculated by using the formula

$$\frac{\text{Weight (kg)} \times \text{Dosage (mg/kg)}}{\text{Concentration (mg/ml)}} = \text{Dose (ml)}$$

The dose thus calculated is slowly injected deep into a large muscle mass (thigh, upper arm, shoulder). Before injecting the tranquilizer, aspiration is performed once to make sure that a blood vessel has not been inadvertently punctured.

Blood is collected under sterile conditions from large, superficial, and easily accessible veins (radial, saphenous, femoral, etc.) into syringes, tubes, or vacutainers, with or without anticoagulant. When anticoagulant is used (collection of blood for hemagglutination test), it must be carefully mixed with blood by rotating the tube or vacutainer. After the vacutainer is filled, care must be taken that air remaining above the blood level is equilibrated with the outside air pressure. If negative air pressure remains in the vacutain-er, it causes hemolysis; if too much blood is forced into the tube so that some air becomes compressed between the blood level and the rubber stopper, this will make the stopper pop up in the air cargo room. Equilibration of the air pressure is best done after the sample is collected in the tube; at that time, the stopper is wiped clean with alcohol or carefully flamed, and it is punc-tured again with a new needle to allow air to flow freely bewteen the vacutainer and the outside. In most cases, for complete blood group testing, 7–15 ml of whole blood in anticoagulant is required. Collected blood specimens are stored at refrigerator temperature (4°C) until ready for shipment or testing.

The anticoagulants most commonly used in primate blood grouping practice are ACD (acid citrate dextrose) or EDTA (ethylenediaminetetraacetic acid). Heparin is less desirable as an anticoagulant agent because heparinized blood cannot be stored for extended periods of time.

Serum of apes and monkeys is tested for the presence of natural antibodies and/or anti-A and anti-B isoagglutinins. Usually, 2–5 ml of serum is required. For this purpose, 5–10 ml of whole blood *without* coagulant is drawn. If the testing (or shipment) of blood is to be delayed, serum is separated by centrifugation or by sedimentation and retraction of the clot, and is frozen or refrigerated immediately after separation.

To obtain *saliva* samples, the anesthetized animal is placed on the abdomen, its mouth is washed clean of food particles and dirt with water, and the free-flowing saliva is collected directly in a tube or aspirated from the oral cavity. To increase salivation, an intramuscular injection of pilocarpine hydrochloride (0.2 mg/kg body weight) is recommended. Generally, 1–3 ml of saliva is required for A-B-H saliva inhibition tests.

The blood group substances present in saliva are quickly destroyed by salivary enzymes. Therefore, immediately after collection, the saliva must be frozen (to slow down the enzymatic activity) and kept frozen until ready for testing. Inactivation of salivary enzymes can also be achieved by placing the tube with saliva, immediately after collection, in boiling water for at least 20 min. Such heat-inactivated saliva does not require freezing and can be stored in a regular refrigerator (at 4°C) for a long period of time.

For *shipment of samples*, vacutainers and tubes have to be carefully labeled (species and identification of animal, date of bleeding), and stoppers, corks, etc., are fastened with adhesive tape. Each tube must be wrapped separately in paper to prevent breakage and to avoid hemolysis by touching the refrigerant.

Blood should be mailed *fresh*. As reported by Rowe et al. [167,168], nonhuman primate red cells have half the life of human erythrocytes and begin to hemolyze after approximately 10 days of storage under refrigeration and much sooner at room temperature. The blood samples must be mailed *under refrigeration but not frozen*. As the coolant, only wet ice or water-ice substitutes are used—such as, for instance, ice-pack bags, artificial cooler ("Scotch") cans, plastic containers, or even frozen beer or soda cans without pop tops. *Dry ice* (CO_2), *is not used*, as it causes freezing and hemolysis of the red cells. The frozen coolant is wrapped in paper towels to prevent direct contact with blood or saliva tubes.

For shipment, insulated containers are used. Empty spaces inside the box are avoided, so that tubes cannot rattle. A duplicate address is placed inside the parcel to avoid loss. Packages are conspicuously marked PERISH-ABLE—FRAGILE—KEEP COLD BUT AVOID FREEZING—RUSH.

If for any reason the blood grouping tests will have to be delayed beyond the normal life-span of the refrigerated red cells, the blood should be frozen, in a special medium, immediately upon collection and held frozen until ready for testing. For details of a simple freezing technique in glycerol-citrate solution consult Erskine and Socha [44]. Methods of freezing red cells in liquid nitrogen are discussed in all their practical aspects by Rowe et al. [166,168–170].

The techniques employed in typing blood of nonhuman primates are essentially the same as, or very similar to, those commonly used in human immunohematolgy, so only those aspects peculiar to blood grouping of primate animals need be discussed here. For basics and details of blood grouping methodology, the reader is referred to the numerous textbooks devoted to theory and practice of human blood groups (see, for example, Bryant [15] and Erskine and Socha [44]).

For purposes of expediency, as well as for historical reasons, discussion of primate blood grouping methodology is best approached from the point of view of the nature of testing reagents.

TESTS WITH REAGENTS ORIGINALLY PREPARED FOR TYPING HUMAN BLOOD

The first blood groups to be investigated in apes and monkeys were the homologues of the human red cell antigens detected by the reagents used for the blood grouping of human blood. Since reagents of this type are by and large readily available commercially and do not require expensive, cumbersome, and time-consuming immunizations of primate animals, the tests for homologues of human blood groups can be performed by any laboratory routinely involved in human grouping work, provided the results obtained are critically evaluated.

Tests for A-B-O Blood Groups

Determination of the A-B-O groups by hemagglutination tests. These determinations are carried out only on the red cells of anthropoid apes. The tests are performed with potent anti-A and anti-B reagents, usually obtained from human donors, and properly standardized for their specificity and titer in control tests with human red cells of various groups [43]. In our practice, no anti-A serum is used for testing primate blood unless it agglutinated human group A_1 red cells at 1:256 dilution, group A_2 cells at a 1:128 dilution, group A_1B cells at 1:128 dilution, and group A_2B cells at a 1:64 dilution. No anti-B serum is used until it yields the titer of 1:128 in tests with human group B cells. The reason for using such high-titer reagents is that they can be diluted to the point at which the nonspecific heteroantibodies against monkey red cells, normally present in human sera, are no longer interfering with the type-specific reactions. Table 9-1 gives an example of comparative titrations of an unabsorbed anti-B serum with red cells of apes and with human erythrocytes of groups B and O. Titrations of this kind are indispensable for establishing the working dilutions of the anti-A and anti-B human sera that are to be used in tests with primate red cells. As can be seen, the serum also contains, in addition to potent anti-B agglutinins, low-titer nonspecific heteroagglutinins that cause agglutination of the red cells of all apes, independent of their A-B-O type. If the serum were to be used undiluted in one-tube tests, it would give positive reactions with all orangutan and gibbon red cells tested, falsely suggesting that all the red cells contained a B agglutinogen. By simply diluting the raw serum with 5–10 volumes of the saline solution, the interfering heteroagglutinins are eliminated and pure anti-B reagent is obtained that is suitable for typing ape red cells.

Elimination of heteroagglutinins by dilution is not practical with antisera of low titers and has the disadvantage of also diluting the group-specific antibody. A more reliable method consists of rendering the anti-A and anti-B sera specific by absorption with chimpanzee group O red cells. The thoroughly washed and packed chimpanzee red cells are mixed with two volumes of the undiluted anti-A or anti-B antiserum and incubated at room temperature for 60 min. After incubation, the tube is centrifuged and the supernatant serum removed immediately and tested with a fresh batch of washed chimpanzee group O cells to control the progress of absorption. If necessary, absorption is repeated until all traces of heteroagglutinins are removed. The resulting reagent can be used full strength for A-B-O typing of chimpanzees, gibbons, orangutans, and gorillas. Specificity of reactions of the reagents thus prepared with the red cells of apes can be ascertained by absorption experiments; for instance, absorption of the anti-A serum with chimpanzee (or gibbon or orangutan) group A red cells renders the reagent inactive against human group A red cells also, while absorption of the serum with human A_1 cells will render the serum inactive against red cells of apes of group A.

TABLE 9-1. Comparative titrations of an unabsorbed anti-B serum of human origin against primate and human red cells suspended in isotonic saline solution

Red cells of:	Dilutions of human anti-B serum										
	1:1	1:2	1:4	1:8	1:16	1:32	1:64	1:128	1:256	1:514	1:1028
Orangutan (group A_1B)	+++	+++	++s	++s	++	++	+±	+	–	–	–
Orangutan (group B)	+++	+++	++s	++s	++	++	++	+s	±	–	–
Orangutan (group A_1)	++w	+±	+	–	–	–	–	–	–	–	–
Gibbon (group A_1B)	+++	+++	+++	++s	++s	++s	++	+±	+	–	–
Gibbon (group A_1B)	+++	+++	+++	++s	++s	++s	++	++w	+s	–	–
Gibbon (group B)	++	+±	+s	–	–	–	–	–	–	–	–
Human group B	+++	+++	+++	++s	++s	++s	++s	++	+s	–	–
Human group O	–	–	–	–	–	–	–	–	–	–	–

Here and throughout the text and tables the strength of agglutination is indicated by the number of plus signs: +++ represents the maximal reaction, namely, one solid aggregate of cells; ++s represents large red clumps, no unagglutinated cells; ++ represents small red clumps, about one-fourth of the cells not agglutinated; ++w represents very small red clumps, still visible microscopically, about one-fourth of the cells not clumped; +± represents clumps clearly visible microscopically, about one half of the cells not clumped; +s represents small clumps visible only microscopically on the background of mainly unagglutinated cells; + represents small clumps visible microscopically on the background of mainly unagglutinated cells; ± represents doubtful reaction, occasional small clumps; – indicates no clumps at all visible microscopically.

Tests for the subgroups of A. These tests are done with anti-A_1 hemagglutinating reagents prepared by absorbing a selected human group B serum with human A_2 cells and also with a reagent prepared by extracting seeds of *Dolichos biflorus*. Reagents of both kinds usually have titers for human A_1 cells of 10—20 units,[2] and slightly lower titers for chimpanzee, gibbon, or orangutan A_1 red cells. Before being used for typing ape blood, the anti-A_1 reagents must be controlled for their specificity by tests on human A_2 cells and chimpanzee group O cells.

Testing of the sera for isoagglutinins anti-A and anti-B. As in human blood grouping practice, the results of tests for A and B agglutinogens on the ape red cells are corroborated by tests on the animal's serum for the presence of anti-A and anti-B agglutinins, which are in most cases of ape blood, reciprocally related to the red cell agglutinogens (Landsteiner's rule). This so-called reverse grouping is done by testing the serum against known human red cells of groups O, A_1, A_2, and B. To eliminate the interfering action of any, weak nonspecific heteroagglutinin that might be present in ape's serum, the serum is either tested diluted or, even better, preabsorbed with human group O red cells. Table 9-2 shows some representative titrations (by saline technique) of the sera of apes with human red cells of various A-B-O types, before and after absorption with human O cells. As can be seen, in most cases the nonspecific heteroagglutinins are so weak that threefold to fourfold dilution of the serum will completely eliminate their interference. Unfortunately, this procedure also considerably weakens the anti-A and anti-B isoagglutinins. More satisfactory results are achieved with undiluted sera absorbed with human group O cells. As shown in Table 9-2, absorption does not significantly reduce the titer and avidity of isoagglutinins for human cells.

When the isoagglutinins are weak and their reactions with human cells hard to evaluate, agglutinations can be enhanced by adding acacia solution to the mixture of serum and the test red cells [242, 256].

Table 9-3 reproduces a typical protocol of tests for the A-B-O blood groups on specimens of various species of anthropoid apes.

Saliva inhibition tests for A, B, and H group substances. This is the main technique for A-B-O testing of the Old and New World monkeys, whose red cells frequently do not react with anti-A , anti-B, or anti-H reagents but who are all secretors of A-B-H substances. The tests are carried out by titration to determine the highest dilution of saliva capable of inhibiting

---

[2]Titer expressed in absolute units is the reciprocal of the highest dilution of serum that gives + (one plus) reaction with positively reacting red cells.

TABLE 9-2. Representative titrations of the normal sera of apes against human red cells of various A-B-O groups

Serum of:	Tested against known human red cells of group:	1:1 (Undil.)	1:2	1:4	1:8	1:16	1:32	1:64
Unabsorbed								
Chimpanzee, group A	O	+		±	−	−	−	−
	A_1	+s		+	−	−	−	−
	A_2	+		±	−	−	−	−
	B	++s		++	+s	±	−	−
Chimpanzee, group O	O	+s		+	−	−	−	−
	A_1	++s		++	++w	+	−	−
	A_2	++		++w	+s	−	−	−
	B	++s		++	++w	+s	−	−
Orangutan, group A	O	++		++	+	−	−	−
	A_1	+s		+	−	−	−	−
	A_2	±		−	−	−	−	−
	B	++s		++s	++	++	+s	+
Gorilla, group B	O	+		−	−	−	−	−
	A_1	++s		++	+±	−	−	−
	A_2	++		+±	±	−	−	−
	B	+		−	−	−	−	−
Absorbed with human O RBCs								
Chimpanzee, group A	O	−		−	−	−	−	−
	A_1	−		−	−	−	−	−
	A_2	−		−	−	−	−	−
	B	++		++w	+	−	−	−
Chimpanzee, group O	O	−		−	−	−	−	−
	A_1	++s		++	+s	+	−	−
	A_2	++w		+s	−	−	−	−
	B	++s		++	+	−	−	−
Orangutan, group A	O	−		−	−	−	−	−
	A_1	−		−	−	−	−	−
	A_2	−		−	−	−	−	−
	B	++s		++s	++s	++	+s	+
Gorilla, group B	O	−		−	−	−	−	−
	A_1	++		++w	+s	±	−	−
	A_2	+s		+s	−	−	−	−
	B	−		−	−	−	−	−

TABLE 9-3. Protocol of tests for A-B-O blood groups in anthropoid apes

Species and animal identification	Reactions of ape red cells with:				Reaction of ape serum[b] with human red cells:				Indicated blood group
	Anti-A serum[a]	Anti-A1 serum[a]	Anti-A1 lectin	Anti-B serum[a]	O	A1	A2	B	
Chimpanzees									
Ch-19 Gabriel	++s	++w	++w	−	−	−	−	++	A_1
Ch-154 Hiram	+++	++w	+±	−	−	−	−	++	A_1
Ch-120 Lolita	++s	+	±	−	−	−	−	++s	$A_{1,2}$
Ch-324 Jackie	++	−	−	−	−	−	−	++	A_2
Ch-168 Walter	−	−	−	−	−	++	++w	++s	O
Gibbons									
G-5 Little	−	−	−	++s	−	++w	±	−	B
G-1 Whitey	++s	++s	++w	++s	−	−	−	−	A_1B
Burma	++s	++s	++s	−	−	−	−	+±	A_1
G-3 Blackey	++s	++s	++	++	−	−	−	−	A_1B
Orangutans									
Bob	++s	−	−	++s	−	−	−	−	A_2B
Maggie	++s	++w	++s	−	−	−	−	+	A_1
Bubbles	++	+s	+w	++s	−	−	−	−	$A_{1,2}B$
Arnold	++	+±	+s	−	−	−	−	++s	$A_{1,2}$
Gorillas									
Binti	−	−	−	+±	−	++	++w	−	B
Dolly	−	−	−	+s	−	++	++	−	B
Jim	−	−	−	+±	−	++w	++w	−	B
Trib	−	−	−	+s	−	++w	++w	++s	B
Human controls									
Group O	−	−	−	−	−	−	−	−	
Group A1	++s	++s	++s	−	−	++	++	++	
Group A2	++	−	−	−	−	−	−	++s	
Group B	−	−	−	++s	−	++s	++	−	

[a] Absorbed with chimpanzee group O cells.
[b] Absorbed with human group O cells.

agglutination of indicator red cells by appropriately prepared anti-A, anti-B, and anti-H reagents.

Three sets of five small (10–75 mm) tubes are prepared, and in each tube of a set is placed one drop of undiluted saliva or 1:4, 1:16, 1:64, or 1:256 diluted saliva. To each tube is then added a drop of appropriate antiserum, and the mixtures are allowed to stand at room temperature for about 30 min. After incubation, a drop of 2% suspension of indicator human red cells of the appropriate blood groups is added. These mixtures are allowed to stand for 60–90 min and are then examined for agglutination, both with the naked eye and under the scanning lens of the microscope.

The anti-A and anti-B sera of human origin to be used in saliva inhibition tests must be titrated and then diluted to yield the titer (at room temperature) of about 8 units for human A_2 cells and B cells. The anti-H reagent is prepared by extracting seeds of *Ulex europaeus* and adjusting its titer to about 4–8 units for human group O cells, at refrigerator temperature.

Saliva inhibition tests are, whenever possible, confirmed by testing the animal's serum for the presence of anti-B and anti-A agglutinins. As described earlier in this chapter, the serum, preabsorbed with human group O red cells to remove traces of nonspecific antibodies, is tested against human A_1, A_2, B, and O red cells suspended in saline.

Table 9-4 shows a typical protocol of A-B-O blood grouping tests on saliva and serum samples from a few representatives of Old and New World monkey species.

Tests for the A-B-O groups in New World monkeys are often complicated by the regular presence on their red cells of a B-like agglutinogen, first demonstrated by Landsteiner and Miller [95]. Their A-B-O blood groups, however, are quite independent of the B-like agglutinogen on the red cells, as can be demonstrated by testing their saliva and serum. For example, all 31 marmosets of various species tested by Wiener et al. [269] were found to be secretors of A and H, with anti-B agglutinins in their serum, while their red cells were agglutinated by anti-B sera. Those observations were independently confirmed and extended by Gengozian [51], who, by using absorption and elution techniques, detected on marmoset red cells an A-like agglutinogen in addition to the B-like agglutinogens.

Table 9-5 shows the composite protocol of hemagglutination tests and saliva inhibition, as well as serum tests carried out with a series of howler monkeys and appropriate controls [49]. As shown in the table, human anti-A and anti-B sera, diluted so as to eliminate interference from nonspecific heteroagglutinins, gave reactions with howler monkey red cells typical of

TABLE 9-4. Example of a protocol of the A-B-O blood grouping tests (quantitative saliva inhibition tests and reverse serum tests) in Old and New World monkeys

Saliva and serum of:		Anti-A serum diluted 1:40 and group A2 red cells; dilutions of saliva:					Anti-B serum diluted 1:5 and group B red cells; dilutions of saliva:				
		1/1	1/4	1/16	1/64	1/256	1/1	1/4	1/16	1/64	1/256
Baboons	1	−	−	−	−	−	++	++	++	++	++
	2	−	−	−	−	+	−	−	−	−	+
	3	++	++	++	++	++	−	−	−	−	−
	4	++	++	++	++	++	−	−	−	+w	+
	5	−	−	−	−	−	−	−	−	−	+w
Gelada		++	++	++	++	++	++	++	++	++	++
Rhesus	1	++	++	++	++	++	−	−	−	−	−
monkeys	2	++	++	++	++	++	−	−	−	−	−
Crab-eating	1	−	−	−	−	−	++	++	++	++	++
macaques	2	−	−	−	−	−	−	−	−	−	+w
	3	++	++	++	++s	++s	−	−	−	+	+w
Pig-tailed	1	++	++	++	++	++s	++	++	++	++s	++s
macaques	2	−	−	−	−	−	++	++	++s	++s	++s
Squirrel	1	−	−	−	−	−	++w	++w	++	++	++s
monkeys	2	−	−	−	−	−	−	−	−	+w	++w
Howler	1	++	++	++s	++s	++s	−	−	−	−	−
monkeys	2	++	++	++	++	++s	−	−	−	−	+w
Human controls[a]											
A, Sec		−	−	−	−	+	++	++	++	++	++
B, Sec		++	++	++	++	++	−	−	−	+	++
O, Sec		++	++	++	++	++	++	++	++	++	++
nS		++	++	++	++	++	++	++	++	++	++

[a]Sec = secretor; nS = nonsecretor.

group B. The importance of the precaution of diluting the antisera to eliminate the effect of heteroagglutinins is evident from the column headed "group AB," which shows that a randomly selected human group AB serum, when used undiluted, strongly agglutinated the red cells of all the howler monkeys as well as other nonhuman primates (and rabbits) due to the presence in the human serum of nonspecific heteroagglutinins for all human red cells. Therefore, to test for the presence of anti-A and anti-B, the sera of howler monkeys were all first absorbed with human group O cells. As shown in Table 9-5, the test then revealed the presence of anti-A in all the sera, thus confirming that all the animals were of group B. Final confirmation of the howler monkeys as group B came from tests on their saliva, which strongly inhibited anti-B and anti-H but failed to inhibit anti-A.

Anti-H lectin (*Ulex*) diluted 1:10 and group O red cells; dilutions of saliva:					Serum test, against human red cells of group:				Human-type A-B-O group
1/1	1/4	1/16	1/64	1/256	O	A₁	A₂	B	
−	−	−	−	−	−	−	−	+++	A
−	−	−	−	−	−	−	−	−	AB
−	−	−	−	+	−	+++	++	−	B
−	−	−	−	−	−	++	++	−	B
−	−	−	−	+	−	−	−	−	AB
−	−	−	−	−	−	−	−	−	O(?)
−	−	−	−	+	−	++	++	−	B
−	−	−	−	−	−	+++	++	−	B
−	−	−	−	−	−	−	−	++	A
−	−	−	−	−	−	−	−	−	AB
−	−	−	−	+	−	+++	++	−	B
−	−	−	−	−	−	++	++w	++	O
−	−	−	−	−	−	−	−	+++	A
−	−	−	−	+	−	−	−	−	A(?)
−	−	−	+w	+	−	−	−	−	AB
−	−	−	+	+s	−	+++	++	−	B
−	−	−	−	+	−	++s	++s	−	B
−	−	−	+w	+					
−	−	−	+w	+					
−	−	−	−	+w					
++	++	++	++	++					

Table 9-5 also shows the reactions of the red cells of the howler monkeys with certain lectins. No agglutination was observed with lectins of *Dolichos biflorus* (anti-A₁), lima bean (anti-A), or *Helix pomatia* snail (anti-A); negative reactions were also obtained with anti-H lectin (*Ulex europaeus*), thus confirming the absence of A and H from howler monkey red cells. Significantly, strong positive reactions were obtained with the lectin of *Evonymus europaeus*, known to have specificity anti-B,H [230].

The results of direct hemagglutination tests would appear to indicate that howler monkeys are indistinguishable from human group B individuals; however, when more refined tests are applied, it becomes obvious that the B agglutinogen on the red cells of these New World monkeys is not identical

TABLE 9-5. Representative results demonstrating the specificities A, B, and H on red cells and in saliva, and anti-A or anti-B agglutinins in sera of 10 mantled howler monkeys (*Alouatta palliata*) among a series of 52 monkeys, with comparative reactions for man, marmoset, rabbit, rhesus, baboon, and gorilla

Red cells of animal: no. and sex	Reactions of red cells with human sera:			Reactions of red cells with lectins		
	Anti-A (dil. 1:10)	Anti-B (dil. 1:10)	Group AB (undiluted)	Anti-H (*Ulex europaeus*)	Anti-A$_1$ (*Dilochos biflorus*)	Anti-A (*lima bean*)
Howler monkeys						
1. No. 32F	−	+ + +	+ + +	−	−	−
2. No. 45M	−	+ + ±	+ + ±	−	−	−
3. No. 48M	−	+ + ±	+ + +	−	−	−
4. No. 49F	−	+ + +	+ + +	−	−	−
5. No. 50F	−	+ + +	+ + ±	−	−	−
6. No. 51F	−	+ + ±	+ + +	−	−	−
7. No. 53F	−	+ + +	+ + ±	−	−	−
8. No. 54F	−	+ + ±	+ + ±	−	−	−
9. No. 55M	−	+ + +	+ + +	−	−	−
10. No. 58M	−	+ + ±	+ + ±	−	−	−
Controls						
Human						
Group O	−	−	−	+ + +	−	−
Group A$_1$	+ + +	−	−	− to + + ±	+ + +	+ + +
Group A$_2$	+ ±	−	−	+ + +	−	+ + ±
Group B	−	+ + +	−	− to + + +	−	−
Marmoset	+ +	+ + +	+ + +	−	−	−
Rabbit	−	+ + +	+ + +	−	−	−
Rhesus	−	−	+ + ±	−	−	−
Baboon group B	−	−	+ + +	−	−	−
Gorilla	−	−	+ + +	−	−	−

with that of the human group B red cells. Table 9-6 presents the results of comparative tests carried out with four anti-B reagents of various origins used in direct tests not only with red cells of human group B, but also with those of howler monkeys, marmosets, and rabbits, all of which are known to have B-like agglutinogen. Red cells of baboons, rhesus monkeys, and gorillas—all group B but with nonreactive red cells—were used as negative controls. As shown in the table, absorption of the two human anti-B sera with human B red cells eliminated the reactivity for human B cells but not for red cells of howler monkeys, marmosets, or rabbits. Absorption with the red cells of either howler monkeys or marmosets removed the reactivity for

Reactions with red cells with lectins		Reactions of serum absorbed with human group O cells with human RBC's of group:				Saliva inhibition titers for:			Indicated human-type A-B-O blood group
Anti-A$_{HP}$ snail (*Helix pomatia*) (diluted)	Anti-B,H (*Evonymus europaeus*)	O	A$_1$	A$_2$	B	A	B	H	
−	+ + +	−	+ + +	+ +	−	0	256	128	B
−	+ + ±	−	+ +	+	±	0	256	256	B
−	+ + +	−	+ + ±	+ +	−	0	256	256	B
−	+ + +	−	+ + +	+ ±	−	0	128	256	B
−	+ + ±	−	+ +	+ ±	−	0	256	256	B
−	+ + +	−	+ + ±	+ +	+	0	256	256	B
−	+ + ±	−	+ + ±	+ +	−	0	256	256	B
−	+ + +	−	+ + ±	+ +	−	0	256	256	B
−	+ + +	−	+ +	+ ±	−	0	256	256	B
−	+ + ±	−	+ +	+ ±	−	0	256	256	B
−	+ + +					0	0	128	
+ + +	− to + +					256	0	256	
+ + ±	+ + +					256	0	256	
−	+ + +					0	256	16	
−	+ + +					256	0	256	
+ +	+ + +					−	−	−	
−	−					0	256	256	
−	−					0	256	256	
−	−					0	128	256	

each other but not for red cells of man or rabbit, while absorption with rabbit red cells removed all the reactivity except for human group B cells. These absorption experiments are quoted here as demonstration of the existence of several fractions of anti-B agglutinins in human anti-B serum: one reactive for human B cells alone, a second reactive for human group B and rabbit red cells but not for monkey red cells, a third reactive for monkey and rabbit red cells but not for human group B red cells, and, perhaps, a fourth that is cross-reactive for all four kinds of red cells having B and B-like agglutino-gens. The results obtained with anti-B of baboon origin were similar, except that with this reagent one of the fractions obtained by absorption differentiated between red cells of howler monkeys and marmosets. The activity of the anti-B,H lectin could be removed completely by absorption with human group B or rabbit red cells. Therefore, in the anti-B,H lectin only two fractions could be isolated by absorption: one cross-reactive with human group B cells and the B-like agglutinogen of the rabbit red cells, and the second cross-reactive with all four kinds of B cells as well as human group O cells.

M-N Typing

For M-N typing, anti-M and anti-N reagents prepared from rabbit antisera are used in hemagglutination tests with primate red cells as in routine human blood typing. Interference from heteroagglutinins is largely avoided by diluting these reagents so that, with some exceptions, they can be used directly and with the same techniques as for man. Anti-N^V lectins (*Vicia graminea* [155] and *Vicia unijuga* [123]) that detect an N-like specificity on human red cells can also be used with good results in tests on primate animals.

Table 9-7 reproduces a protocol of tests carried out with red cells of various primate species, using anti-M and anti-N reagents of different origin [254]. As can be seen, the tests are performed in parallel with batteries of various reagents of the same specificity, since some of the reagents may fail to detect fractions of M and N antigens present on the red cells of various nonhuman primate species and thus give misleading results if used in single-reagent tests. In cases of weak reactions, when there may be some uncertainty as to whether the factor in question (for example, M) is really present on the ape or monkey red cells, the reagent is absorbed with positively reacting human blood. This should eliminate the reactivity for ape red cells also if the weak reactions with unabsorbed anti-M reagents were actually due to an M specificity of ape red cells and not to the nonspecific heteroagglutinins in rabbit serum. On the other hand, absorptions with positively reacting primate red cells can be used for fractionation of anti-M or anti-N reagents that are known to be mixtures of antibodies directed against various parts of the very complex M or N antigens. Table 9-8 shows the results of one such absorption-fractional experiment described by Wiener et al. [254].

Rh-Hr Typing

For these tests, Rh-Hr antisera of human origin are used. The same techniques are applied in tests with red cells of primate animals as for testing human red cells, depending on the requirements of the given reagent. By and large, the ficinated red cell technique gave, in our hands, dependable results with most of the reagents and with the least effort.

TABLE 9-6. Demonstration by absorption of the differences among the B agglutinogens on the red cells of man, howler monkeys, marmosets, and rabbits

Anti-B reagent and absorbing red blood cells	Tested against red blood cells of:								
	Human			Howler monkey	Marmoset	Rabbit	Baboon group B	Rhesus	Gorilla
	O	A	B						
Anti-B commercial (1:10)									
Unabsorbed	—	—	++	++s	++	++	—	—	—
Absorbed with human B	—	—	—	++s	++	+	—	—	—
with howler monkey	—	—	++s[a]	—	—	++w[a]	—	—	—
with marmoset	—	—	++s	—	—	++w	—	—	—
with rabbit	—	—	++w	—	—	—	—	—	—
Anti-B (Leff) (undiluted)									
Unabsorbed	—	—	++	++s	++s	++	—	—	—
Absorbed with human B	—	—	—	++	++	++s	—	—	—
with howler monkey	—	—	++s	—	—	+++	—	—	—
with marmoset	—	—	+++	—	—	+++	—	—	—
with rabbit	—	—	++w	—	—	—	—	—	—
Anti-B,H lectin[b] (undiluted)									
Unabsorbed	++w	—	++	++s	++s	++w	—	—	—
Absorbed with human B	—	—	—	—	—	—	—	—	—
with howler monkey	—	—	++w	—	—	++s	—	—	—
with marmoset	—	—	++w	—	—	++s	—	—	—
with rabbit	—	—	—	—	—	—	—	—	—
Anti-B baboon[c] (undiluted)									
Unabsorbed	—	—	+++	+++	+++	++s	—	—	—
Absorbed with human B	—	—	—	++s	++s	++s	—	—	—
with howler monkey	—	—	++s	—	+s	++s	—	—	—
with marmoset	—	—	++s	—	—	++s	—	—	—
with rabbit	—	—	—	—	+s	—	—	—	—

[a] s = strong; w = weak.
[b] *Evonymus europaeus* lectin.
[c] Actually preabsorbed with human group A red blood cells to remove nonspecific heteroagglutinins.

TABLE 9-7. Results of tests on red cells of gorillas, orangutans, and other simians, with anti-M, anti-Mᵉ, anti-He, anti-Nᵛ reagents

Tested against red cells of:	Anti-M (anti-human)			Anti-M (anti-chimpanzee)	Anti-M (anti-baboon)	Anti-He (anti-human)	Anti-Mᶜ (anti-human)	Anti-Nᵛ lectin (Vicia graminea)
	M1	M2	M6					
Gorillas								
1. Anka	+++	tr	–	+++	+++	+++	++	+++
2. Banga	+++	±	+	+++	+++	+++	+±	+++
3. Choomba	±	–	–	–	+±	++	+	+++
4. Calabar	+++	–	–	+++	+±	–	–	+++
5. Chad	tr	–	–	–	tr	+++	+±	+++
6. Inaki	+++	+±	+±	+++	++	+++±	+±	+++
7. Jini	+++	+	tr	+++	+++	+++±	+±	+++
8. Katoomba	+++	–	+	+++	+++	+++±	+	+++
9. Oban	+++	+±	–	+++	++	–	–	+++
10. Oko	+++	++	++	+++	+++	+++	++	+++
11. Ozoom	–	–	–	–	–	+++	++	+++
12. Paki	+++	++	+++	+++	+++	++	+±	+++
13. Rann	±	–	–	–	+	++	+±	+++
14. Shamba	+++	+±	++	+++	+++	++	+	+++
15. Segon	+++	–	tr	+++	++	+±	+±	+++
Human controls								
16. Type M, He–	+++	+++	+++±	+++	+++	–	++	–
17. Type N, He–	–	–	–	–	–	–	–	+++
18. Type MN, He–	+++	++±	+++	+++	+++	–	++	+++
19. Type N, He+	–	–	–	–	–	+++	+++	+++

Orangutans								
20. Paddi	(++++)	(++)	(++)	(++++)	(++++)	(++)	(++)	—
21. Tupa	(++)	(++)	(++)	(++)	(+++±)	(+++±)	(+++±)	—
Chimpanzees								
22–26. 5 animals	+++	+++±	+++±	+++	+++	—	++	nd
Gibbons								
27. G-12 Abby	+++	nd	+±	+++	++±	—	tr	nd
28. G-3 Blackey	+++	nd	—	++	+	—	—	nd
Baboons								
29–31. 3 animals	+++	++	++	+++	+++	—	—	nd
Rhesus monkeys								
32–33. 2 animals	+++	nd	+±	+++	+++	—	—	nd
Crab-eating macaques								
34–35. 2 animals	+++	nd	+	+++	+++	—	—	nd
Geladas								
36–37. 2 animals	+++	nd	++	+++	+++	—	—	nd
Celebes black apes								
38–39. 2 animals	+++	++	++	+++	+++	—	—	nd
Capuchin monkey								
40. C4 Leah	—	—	—	—	±	—	—	nd

nd = not done; tr = trace.

Parentheses mean that the reaction was not removed by absorbing the reagents with red cells of the type used for immunizing the rabbit.

From Wiener et al. [254].

TABLE 9-8. Fractionation by absorption of the anti-M reagent (M_1) prepared from an immune rabbit serum for human red cells

Tested against red cells of:	Unabsorbed	Absorbed with red cells of:							
		Human M	Gorilla Choomba	Gorilla Katoomba	Chimpanzee	Orangutan Paddi	Baboon	Crab-eating macaque	Celebes black ape
Gorillas									
1. Anka	+++	−	+++	−	−	−	−	−	−
2. Banga	+++	−	+++	−	−	++	−	−	−
3. Choomba	−	−	−	−	−	−	−	−	−
4. Calabar	+++	−	++	−	−	−	−	−	−
5. Chad	−	−	−	−	−	−	−	−	−
6. Katoomba	+++	−	++±	−	−	−	−	−	−
7. Oban	+++	−	++±	−	−	+±	−	−	−
8. Chimpanzee #335	+++	−	++	−	−	+±	−	−	−
Orangutans									
9. Paddi	+++	++	++±	++	+±	+±	++	++	++
10. Tupa	++	++	++±	++	+±	+±	+±	++	++
Human controls									
11. Type M	+++	−	++±	++±	+++	+++	+++	+++	++±
12. Type N	−	−	−	−	−	−	−	−	−
13. Type MN	+++	−	+++	++	+++	++±	+++	+++	+++
14. Baboon #194	+++	−	++±	−	−	±	−	−	−
15. Crab-eating macaque	+++	−	++±	−	−	−	−	−	−
16. Rhesus monkey #55	+++	−	++±	−	−	±	−	−	−
17. Celebes black ape #13	+++	−	++±	−	−	−	−	−	−
18. Capuchin #4	−	−	−	−	−	−	−	−	−

Further tests are necessary to determine whether the reagents contain nonspecific heteroagglutinins for ape or monkey blood; no specific method has been devised to remove them without weakening the type-specific antibodies.

When the reagents give negative reactions in direct tests with primate animal blood, the specificity being tested for is assumed to be absent from the tested cells. It must be stressed, however, that more than one serum of a given specificity may be necessary for side-by-side testing with primate red cells, since—as is also the case with anti-M and anti-N reagents—Rh-Hr reagents may differ in their reactivity with ape and monkey blood, though they give identical or comparable results with all positively reacting human red cells.

If positive reactions are obtained, further tests are required to determine whether they are type-specific or due to other antibodies in the reagent. One of the methods used is titration of the positively reacting simian red cells in parallel with positive human red cells. Since the reagents are in general of high titer, similarity of titers may be considered evidence that the reactions are indeed type-specific. Tables 9-9 and 9-10 give examples of such comparative titrations of the Rh-Hr reagents as reported by Wiener in one of his early works on chimpanzee blood groups [251]. Similarities of titers lead to the conclusion that chimpanzee red cells share with human Rh-positive erythrocytes the Rh_0 (D) and hr' (c) specificities, but not rh', rh'', Hr, or hr'' specificities. In another, similar experiment, comparative titrations of anti-hr' (c) antisera (see Table 9-11)) were carried out with orangutan, chimpanzee, gibbon, and human red cells, showing that, contrary to earlier claims, orangutans are essentially hr' (c)-negative [140].

A more definitive proof of specificity of reactions with Rh-Hr sera is obtained by absorption experiments. If the reagent (anti-Rh_0, for example), after absorption with human Rh_0-positive red cells, becomes inactive also against ape red cells, the reaction of unabsorbed serum with ape red cells is considered type-specific. As is the case with anti-M and anti-N sera, human Rh-Hr reagents can be fractionated by absorption with ape red cells. Table 9-12 gives examples of such absorptions, fractionations, and comparative titrations of anti-Rh_0 sera using human and chimpanzee red cells. The experiments not only confirmed type specificity of reactions with ape red cells, but also proved the existence of fractions in human anti-Rh_0 sera that detected specificities of the simian type (R^c, originally called L^c, and c^c) on the red cells of chimpanzees [199].

TABLE 9-9. Results of tests with Rh antisera on blood specimens of 14 chimpanzees

Blood of:	Anti-rhesus guinea pig serum No. 1, saline (titer)	Anti-Rh₀ (anti-D) Serum No. 2, saline (titer)* a	b	Serum No. 3, ficin (titer)	Serum No. 4, ficin (direct)	Anti-rh' (anti-C) Serum No. 5, blocked anti-Rh₀', saline (direct)	Serum No. 6, ficin (titer)	Anti-rh" (anti-E) Serum No. 7, blocked anti-Rh₀", saline (direct)	Serum No. 8, ficin (titer)	Anti-Rhʷ1 (anti-Cʷ) Serum No. 9, anti-globulin (direct)	Rh type of chimpanzee
Chimpanzee											
1. Iris	3	44	24	180	+++	tr	0	–	0	–	Rh_0^{Ch}
2. Evelyn	2	24	28	190	+++	tr	0	–	0	–	Rh_0^{Ch}
3. Sampson	3	44	32	220	+++	–	0	–	0	–	Rh_0^{Ch}
4. Jerry	2	32	32	180	+++	–	0	–	0	–	Rh_0^{Ch}
5. Cindy	3	24	28	180	+++	–	0	–	0	–	Rh_0^{Ch}
6. Queenie	3	40	48	250	+++	–	0	–	0	–	Rh_0^{Ch}
7. Bike	1	24	24	190	+++	tr	0	–	0	–	Rh_0^{Ch}
8. Patch	2	24	24	180	+++	tr	0	–	0	–	Rh_0^{Ch}
9. Tiny	5	16	20	250	+++	tr	0	±	0	–	Rh_0^{Ch}
10. Lady	2	24	20	180	+++	–	0	–	0	–	Rh_0^{Ch}
11. Georgie	2	16	24	220	+++	±	0	±	0	–	Rh_0^{Ch}
12. Baldie	2	24	48	220	+++	±	0	±	0	–	Rh_0^{Ch}
13. Dotty	2	20	30	220	+++	+	0	±	0	–	Rh_0^{Ch}
14. J.B.	2	14	12	180	+++	–	0	tr	0	–	Rh_0^{Ch}
Controls (human cells)											
Type rh	2	0	0	0	–	–	0	–	0	–	
Type $Rh_1^wRh_1$	24	48	40	160	+++	+++	200	–	0	+++	
Type Rh_2rh	12	24	14		+++	–	0	+++	90	–	
Type Rh_2Rh_0	16	14	12	140	++±	++±	190	++	100	–	

*a = reagent diluted in saline; b = reagent diluted in saline and 30% bovine albumin.

TABLE 9-10. Results of tests with Hr antisera on blood specimens of 14 chimpanzees

Blood of:	Anti-hr' (anti-c)		Serum No. 12, ficin (titer)	Anti-hr" (anti-e) serum No. 13, ficin (titer)	Anti-hr" + anti-rh'(anti-C+e) serum No. 14, ficin (titer)	Anti-Hr + anti-hr^s (anti-C+c+E+e+e^s) serum No. 15, ficin (titer)	Rh-Hr type of chimpanzee
	Serum No. 10, ficin (direct)	Serum No. 11, ficin (titer)					
Chimpanzee							
1. Iris	+ + ±	21	12	0	0	0	$\overline{Rh}_0^{Ch}$ (D^{u}c$-^{ch}$)
2. Evelyn	+ + ±	18	4	½	0		$\overline{Rh}_0^{Ch}$
3. Sampson	+ + ±	18	11	½	0		$\overline{Rh}_0^{Ch}$
4. Jerry	+ + ±	36	20	0	0		$\overline{Rh}_0^{Ch}$
5. Cindy	+ + ±	9	6	0	0		$\overline{Rh}_0^{Ch}$
6. Queenie	+ + ±	33	10	0	0		$\overline{Rh}_0^{Ch}$
7. Bike	+ + ±	21	11	0	0		$\overline{Rh}_0^{Ch}$
8. Patch	+ + ±	18	6	0	0		$\overline{Rh}_0^{Ch}$
9. Tiny	+ + ±	72	12	0	0		$\overline{Rh}_0^{Ch}$
10. Lady	+ + ±	33	14	0	0		$\overline{Rh}_0^{Ch}$
11. Georgie	+ + ±	15	12	0	0		$\overline{Rh}_0^{Ch}$
12. Baldie	+ + ±	30	12	0	0		$\overline{Rh}_0^{Ch}$
13. Dotty	+ + ±	36	11	0	0		$\overline{Rh}_0^{Ch}$
14. J.B.	+ + ±	24	9	0	0	0	$\overline{Rh}_0^{Ch}$
Controls (human cells)							
Type rh	+ + +	40	16	26	48	12	
Type Rh$_1$Rh$_1$	−	0	0	24	0	24	
Type Rh$_2$Rh$_2$	+ + ±	36	12	0		28	
Type Rh$_2$rh	+ + ±	30	16	12	24	30	
Type Rh$_2$Rh$_0$							

TABLE 9-11. Comparison of orangutan, chimpanzee, and gibbon red cells in tests with human anti-hr' serum

	Findings from previous study[a]			New findings	
	Average titers with human anti-hr' sera			Average titers with human anti-hr' sera	
Red cells of:	Serum No. 6	Serum No. 10	Red cells of:	Serum Rin.	Serum Gav.
Orangutan			Orangutan		
Lipis	6	12	Lemback	0	0
Kuala	7	4	Padang	1	0
Sampit	6	12	Bagan	1	0
Kitchie	6	10	Bukit	1	0
Lemback	2	2	Ini	1	0
Sya	6	8	Dinding	1	0
Djambi	3	3			
Chimpanzee			Chimpanzee		
Jenda	48	44	Jaylen	256	32
Boka	28	40	Sean	300	nd
Gibbon					
Becky	96	44			
Man			Man		
Type Rh_1rh	56	96	Type Rh_1rh	512	96
Type Rh_1Rh_1	0	0	Type Rh_1Rh_1	0	0
Type rh	96	96	Type rh	512	96

nd = not done.
[a]Data from Wiener et al [267a].

Lewis Typing

Testing of red cells of nonhuman primates for the Lewis types has not been possible because of the lack of potent antisera free of nonspecific heteroagglutinins. However, as tests on the secretions are carried out by the inhibition technique, the presence of heteroagglutinins does not interfere with those tests. Saliva inhibition tests for the presence of the Lewis substance are performed, as in the case of A, B, and H substances, by quantitative techniques. Series of dilutions of the saliva are incubated at body temperature for 1 h with anti-Lewis serum from human donors lacking Lewis substances in their salivas, and then the ficinated indicator red cells from a person of type O, Le(a+), nS are added and reactions are observed in all tubes. Enzyme treatment of the indicator red cells is done as follows: The red cells are washed twice with saline solution, and then nine drops of the packed, washed cells are mixed with one drop of a 1% freshly prepared saline solution

TABLE 9-12. Results of comparative titration and absorption experiments with human anti-Rh$_0$ serum

| | | | Titer by ficinated red cell method | | |
| | | | After absorption with red cells | | |
Tested with red cells of:		Unabsorbed	Pooled chimpanzee R^c-negative	Pooled chimpanzee R^c-positive	Pooled human Rh-positive
Chimpanzees					
Mack	(R^c-positive)	640	40	0	0
Berry	(R^c-positive)	960	40	0	0
Champ	(R^c-positive)	960	40	0	0
King	(R^c-negative)	640	30	0	0
Stan	(R^c-negative)	500	0	0	0
Sean		500	0	0	0
Human					
Roman F.	(Rh$_1$rh)	960	120	120	0
Frank S.	(Rh$_1$Rh$_2$)	960	120	120	0
M.C.	(Rh$_2$Rh$_2$)	960	240	240	0
1382–4	(Rh$_0$)	900	100	120	0
1382–5	(rh'rh)	0	0	0	0
James S.	(rh"rh)	0	0	0	0
138208	(rh)	0	0	0	0

of powdered ficin. After incubation for 1 h at 37°C, the action of the enzyme is stopped by washing the cells and a 2% suspension of red cells is prepared for tests.

Table 9-13 reproduces results of comparative inhibition tests for substances H and Le in the saliva of various species of primates [253]. As can be seen, there is no difficulty in distinguishing between Lewis-positive and Lewis-negative as well as between secretor and nonsecretor types in apes and monkeys.

Testing for I and i Blood Factors

The limited tests for I and i specificities on red cells of apes and monkeys were carried out by Wiener et al. [272] using a cold agglutinating anti-I serum of extraordinarily high titer obtained from a woman with severe autohemolytic anemia [289]. The anti-i serum, also of very high titer, was derived from a man with hemolytic anemia of unknown origin. The tests with both reagents were done by titration in saline medium, at room temperature. Because of the high titers of both sera, the titrations were started with the serum previously diluted 15–20 times, which precluded any interference

TABLE 9-13. Results of inhibition tests for substances H and Le in the saliva of primates

Saliva of:	Reactions of anti-H lectin[a] with group O cells after addition of saliva:					Reactions of anti-Le serum with fucinated Le(a+) red cells after addition of saliva:					Indicated saliva type[b]
	Undil.	1:4	1:16	1:64	1:256	Undil.	1:4	1:16	1:64	1:256	
Chimpanzees											
Pan troglodytes											
Big Bonnie	−	−	++	++	++	+±	++	++			Sec nL
Bogan	−	−	−	+	++	+±	+±	++			Sec nL
Peck	−	−	tr	++	++	+±	++	++			Sec nL
Polly	−	−	tr	+	++	−	tr	+±	++		Sec Les (weak)
Sue	−	−	tr	+	+±	++	++	++			Sec nL
Orangutans											
Pongo pygmaeus											
0-18	−	−	−	+±	++	−	−	−	−	+	Sec Les
0-24	−	−	−	+	++	−	−	tr	+	+±	Sec Les
0-22	−	−	+±	++	++±	−	−	tr	+	++	Sec Les
Sabu	++	++	++±			−	−	−	−	tr	nS Les
Gibbons											
Hylobates lar											
Charlie	−	−	−	+	++	+±	++	++			Sec nL
Viet	−	−	+	+±	+++±	++	++	++			Sec nL
Thai	−	−	−	+±	+++±	+±	++	++			Sec nL

Pigtail macaque
Macaca nemestrina

Cebus monkey
Cebus albifrans

Human controls

Nanda	–	–	–	tr	+±	–	–	–	+±	++	Sec Les
Oakie	–	–	–	–	+	–	–	–	+	+±	Sec Les
Kate	–	–	–	–	tr	–	–	–	+±	++	Sec Les
No. 1	–	–	–	tr	+	+±	++	–	–	–	Sec nL
No. 2	–	–	–	tr	+±	++	++	–	–	–	Sec nL
No. 3	–	–	–	–	–	+±	++	–	–	–	Sec nL
Sec nL	–	++	+++±	+++±	–	++	++	–	–	–	
nS Les	+++±	+++	+++	+++±	–	–	–	–	–	tr	
Sec Les	–	++	++	+++±	–	–	–	–	tr	+	

[a] Saline extract of *Ulex europaeus* seeds.

[b] Sec nL = secretor of H substance, nonsecretor of Lewis substance; Sec Les = secretor of H and Lewis substances; nS Les = nonsecretor of H substance, secretor of Lewis substance; tr = trace.

from heteroagglutinins the sera might have contained. The patient from whom the anti-i serum was obtained belonged to group O, but the high dilution at which the serum was used largely excluded the possibility of interference from the isoagglutinins anti-A and anti-B. In doubtful cases, however, the tests were to be repeated after serum had been absorbed with red cells of human adults of groups A_1 and B. Table 9-14 gives a composite protocol of tests carried out by Wiener et al. [272] with anti-I and anti-i reagents on red cells of various primates. For comparative purposes, the tests were conducted in parallel with the cells of various domestic animals, as well as with human adult and cord cells.

Tests for Homologues of Other Human Bood Groups

The reagents for testing most other human blood groups are generally of low titer; therefore, the results are poorly reproducible even in man, and the reagents are unsuitable for testing red cells of apes and monkeys because of interference of nonspecific heteroagglutinins, which cannot be eliminated without weakening the specific antibodies to the point of inactivity. In spite of these limitations, there are reports of successful attempts to detect the Duffy-related specificities on the red cells of apes, Old World monkeys and even New World monkeys, either by means of direct agglutination or by an absorption-elution method, using specific antisera of human origin [118a,156a]. While anti-Fy^a failed to react with any of the primate animal red cells, the anti-Fy^b sera gave clearly positive reactions with red cells of several nonhuman primate species (chimpanzee, gibbon, rhesus monkey, the crab-eating, stump-tailed, and pig-tailed macaques, patas monkey, and guenon). The same reagents reacted only weakly with blood of gorilla and aotus. The anti-Fy3 antisera were found to agglutinate the red cells of the great majority of primate species tested. In limited experiments with anti-Fy5 and anti-Fs, only the red cells of mountain gorillas reacted with both of these reagents, while the blood of one of the rhesus monkeys tested was agglutinated exclusively by the latter reagent.

TESTS WITH REAGENTS PRODUCED BY IMMUNIZATION OF PRIMATE ANIMALS

Despite early claims of the impossibility of immunizing primates with the red cells of animals of the same or closely related species, isoimmunization and cross-immunization have become the most convenient and widely used ways of producing antisera that recognize monkeys-own red cell specificities.

Immunization

In most of our experiments, the iso- or cross-immune antisera have been produced by *intramuscular injections* of red cells with addition of appropriate vehicle that enhances the antigenicity of blood cells. The best results are achieved with a water-in-oil emulsion in which antigen solution is emulsified in mineral oil (*Freund's incomplete adjuvant*), sometimes with the inclusion of killed mycobacteria (*Freund's complete adjuvant*) to further enhance antigenicity.

To prepare inoculum, mix 0.5 ml of washed, packed donor red cells with 0.5 ml of commercially available Freund's (complete or incomplete) adjuvant (Difco Laboratories, Detroit, Michigan) in a 6 ml sterile syringe. Put the syringe containing the mixture in a shaker and shake for at least 15 min to achieve adequate emulsification. Test the suspension by expressing one drop into a beaker of water. If mixing is adequate, the droplet will float as an intact globular bead on the surface of the water. If the suspension does not pass the droplet test, continue shaking.

The mixture of red cells and adjuvant is injected intramuscularly into multiple sites. Use as many different sites as necessary to inject the full amount of inoculum, depositing 0.2 ml or less into each site. Preferable sites of injections are in the gluteal or lower back region, on ventral or dorsal surfaces of the upper parts of arms, and on ventral surfaces of the upper parts of legs. Injections should be deeply intramuscular. The utmost care should be taken to avoid injecting close to nerves and vessels. This means that inoculum should be deposited in the middle of the muscle mass, and not close to bone or near fasciae and tendons.

The intramuscular injection is followed by a test bleeding after 4–6 weeks, and, if necessary, injections are repeated until antibodies of adequate titer appear in the serum. In most cases, the first injection is carried out with red cells mixed with *complete* Freund's adjuvant, while the following injections are done using a mixture of red cells with *incomplete* adjuvant. It must be recalled that some objections were raised against the use of complete Freund's adjuvant in primate colonies where tuberculin tests are routinely applied as an epidemiological safeguard. Positive hypersensitivity reactions appearing in animals injected with complete adjuvant may cause confusion as to the diagnostic meaning of the results of the test. As shown, however, in past extensive immunization programs, proper management of tuberculin-positive animals made that objection irrelevant, particularly in view of the advantages of this immunization technique: good immunological response of the recipient animals as well as long-lasting, high titers of the produced antibodies.

Alternative methods of immunization call for *intravenous* administration of the antigen. In this technique, animals receive intravenously 10–25 ml

TABLE 9-14. The blood factors I and i in infrahuman species

Blood of:	Undil.	1:2	1:4	1:8	1:16	1:32	1:64	1:128	1:256
Reactions with anti-I serum									
Chimpanzee (1) *Pan troglodytes*	–	–	–						
Gibbon *Hylobates lar lar* (5)	–	–	–						
Orangutan (1) *Pongo pygmaeus*	–	–	–						
Yellow baboon (1) *Papio cyncocephalus*	–	–	–						
Gelada (1) *Theropithecus gelada*	–	–	–						
Crab-eating macaque (1) *Macaca fascicularis*	–	–	–						
Stump-tailed macaque (1) *Macaca arctoides*	–	–	–						
Rhesus monkey (6) *Macaca mulatta*	–	–	–						
Pig-tailed macaque (1) *Macaca nemestrina*	–	–	–						
Sooty mangabey (1) *Cercocebus atus*									
Black and white capuchin (1) *Cebus capucinus*	+++	+++	+++	+++	+++±	+++	++	++	++
Squirrel monkey (1) *Saimiri sciurea*	–	–	–						
Black spider monkey (1) *Ateles paniscus*	++	±	+	–	–	–			

Species									
White-whiskered spider monkey (1)									
Ateles marginatus	++	+±	–	–	–	–		+++	+++	++
Sheep (1)	–	–	–	–	–	–	–	–		
American white-tailed deer (1)	–	–	–	–	–	–	–			
Goat (1)	–	–	–	–	–	–	–			
Cat (1)	–	–	–	–	–	–	–			
Dog (1)	–	–	–	–	–	–	–			
Rabbit (2)	+++	+++	+++	+++	+++	+++	+++	+++	+++	++
Rat (1)	–	–	–	–	–	–	–			
South American agouti (1)	–	–	–	–	–					
Controls										
Human adult	++±	++	++	++	+	+	+		–	
Human newborn (cord)	–	–	–	–	–	–	–		–	
Reactions with anti-i serum										
Chimpanzee (1) *Pan troglodytes*	+++	+++	++±	+++	++	+±	±| +++		–
Gibbon *Hylobates lar lar* (5)	–	–	–	–	–	±	–			
Orangutan (1) *Pongo pygmaeus*	++	+±	–	–	–	–	–			
Yellow baboon (1) *Papio cynocephalus*	+++	+++	+++	++±	++	++	+±		–	
Gelada (1) *Theropithecus gelada*	+++	+++	+++	++	++	++	+±	+	+	–
Crab-eating macaque (1) *Macaca fascicularis*	+++	+++	+++	++±	++	++	++	++±	+±	–
Stump-tailed macaque (1) *Macaca arctoides*	+++	+++	+++	+++	++±	+++	++	++	++	tr
Rhesus monkey (6) *Macaca mulatta*	+++	+++	+++	++±	++	++	++	++	++	+

(Continued on next page)

TABLE 9-14 (Continued)

Blood of:	Undil.	1:2	1:4	1:8	1:16	1:32	1:64	1:128	1:256
Pig-tailed macaque (1)									
Macaca nemestrina	+++	+++	+++	+++	+++±	++	+±	tr	—
Sooty mangabey (1)									
Cercocebus atus	+++	+++	+++	+++	++	++	++	—	—
Black and white capuchin (1)									
Cebus capucinus	+±	+	—	—	—				
Squirrel monkey (1)									
Saimiri sciurea	+++	++	++	+±	+±	+±	—	—	—
Black spider monkey (1)									
Ateles paniscus	+++	+++	++	+±	+	+±	—	—	—
White whiskered spider monkey (1)									
Ateles marginatus	+++	+++	+++	+++	+++±	++	++	+	
Sheep (1)	—	—	—						
American white-tailed deer (1)	—	—	—						
Goat (1)	—	—	—						
Cat (1)	—	—	—						
Dog (1)	++	++	+	+±			—		
Rabbit (2)	—	—	—	—					
Rat (1)	+++±	++	+±	tr					
South American agouti (1)	—	—	—	—					
Controls									
Human adult	+	—	—	—	—				
Human newborn (cord)	+++	+++±	+++±	++	++	+±	+±	—	—

Numbers in parentheses represent the number of animals of each species examined; tr = trace.

(depending on size of the recipient) of 50% suspension of red cells (carefully washed and resuspended in saline or in any intravenous solution). Since antibodies appear early following the intravenous administration of the antigen and the rise of the titer is, by and large, steep and the peak quickly receding, timing of test bleedings is of great importance. In most cases, the highest titers are observed between 7 and 10 days after injection, so it is recommended that the test bleeding be done a week after injection, and if the adequate titer is observed, harvesting of the serum should follow immediately thereafter.

In selected cases, excellent results were obtained by *intraperitoneal* injections of red cells. Depending on the weight of the recipient animals, between 10 and 100 ml of washed, resuspended red cells were administered into the peritoneal cavity. Test bleeding usually followed 1 week later. In general, if a positive response was obtained, the titers of antibodies reached their peak by that time and stayed unchanged for a few weeks. In rare cases, particularly when intraperitoneal injection was used as a booster that followed another technique of immunization, very high titers were attained shortly after the red cells were deposited intraperitoneally, remaining at that level for several months, and weekly plasmaphereses yielded impressive quantities of very good antiserum. It is needless to stress that whenever intraperitoneal injections are performed, necessary precautions must be taken to avoid the possibility of infection or injury to the abdominal organs, which may result in massive hemorrhage or peritonitis, often fatal to the animal.

No definitive scheme of immunization procedure can be offered, since the progress of immunization depends upon many factors, among which the immunogenic properties of the given antigen as well as the immunological constitution of the recipient animal are of crucial importance. While in some animals one or two consecutive injections proved sufficient to produce the desired effect, in some others months and sometimes years of repeated boosting injections were needed to complete the immunization program. In such cases, alternation of immunization techniques and periods of rest between immunization courses were often of some help.

Although, in general, the anthropoid apes seemed to be more responsive to isoimmunization than Old World monkeys, individual differences among animals of the same species were sometimes more accentuated than those among various species of primates. As a general rule, the red cells obtained from an animal belonging to the same species as the recipient are less immunogenic than red cells from animals of different but closely related species; isoimmunization, however, is more practical than cross-immunization, because it results more often in production of monospecific reagents, or the resulting antisera require less absorptions.

Testing of Antisera

To follow the progress of immunization, small quantities of blood are collected from immunized animals immediately before the first course of immunization (preimmunization control), and then at intervals dictated by the choice of immunization technique. Serum (or plasma) is separated from red cells and tested by titration against the donor's red cells and the recipient's own cells (negative control).

The methods of testing and titrating are the same as in human blood typing: the saline agglutination method, the antiglobulin method, the ficinated red cell method, and at times the ficinated red cell antiglobulin method. Other workers have tested rhesus monkey red cells by the dextran method [64, 100,101] originally developed for tests on mice, for which species the methods used for man are inapplicable. Since the objective of blood grouping studies in nonhuman primates has been to approximate conditions in man, it seemed inappropriate to use a method developed for mice, especially since the technique for human blood typing, proved in millions of tests, gives clear, reproducible results with the blood of nonhuman primates. Moreover, in our hands the dextran method does not always give clear-cut and reliable reactions.

A microtiter hemagglutination technique devised by Wegman and Smithies (quoted by [223]) for typing human erythrocytes with commercial typing sera at unusually high dilutions was adapted by Sullivan et al. for use with rhesus alloimmune sera [223]. In this method, rhesus antisera diluted with buffered saline-polyvinylpyrrolidine (PVP)-Tween solution are mixed with highly diluted suspensions of red cells to which normal rhesus serum has been added. The agglutination tests are carried out in rigid polystyrene microtiter plates and results are read with the naked eye after 15 min incubation at 37°C and short centrifugation. According to the authors, the great advantage of the method is its ability to enhance the titer of the reagents, which therefore can be used highly diluted, with great saving of typing reagents. The technique is characterized as much more sensitive than a comparable tube test, thus giving a greater number of clearly positive results. The practical value of the method is considerably limited, however, by the fact that it is suitable only for use with saline-agglutinating antibodies and cannot be applied by antiglobulin technique. As we know, most of the rhesus isoimmune antisera are of the "univalent" type and only rarely contain saline agglutinating fractions. The latter, if present, are of very low titer and avidity. Also, for some unknown reason, some of the saline-agglutinating reagents tested by Sullivan et al. failed to react by microtiter technique, though they were normally active when used by the tube method.

Each serum sample is tested in parallel by three methods—saline agglutin- ation, antiglobulin, and ficin—because different antibodies react by different techniques. In Table 9-15 are shown, for example, comparative titrations of two isoimmune rhesus monkey sera against red cells of the donor monkey, using the saline agglutination, antiglobulin, ficinated red cell, and dextran methods. As can be seen, the antibodies in either serum are of the "univa- lent" or "incomplete" variety, since no clumping occurred by the saline method using untreated red cells. The presence of high-titered antibodies is nevertheless demonstrated by the antiglobulin test. By the ficin method, the first serum gave good reactions but of lower titer than by antiglobulin, while the second serum reacted much more weakly and with a prozone. The results by the dextran method were the least distinct.

Production of Antiglobulin Sera

Theoretically, the antiglobulin used for the tests should be prepared by immunization with the serum or globulin of the species being studied. However, as practice has shown, antiglobulins prepared against one species of monkey can be used in tests with the red cells of all species of Old World monkey, while antiglobulin produced with human globulin gives virtually the same results with human red cells as with the cells of anthropoid apes and only slightly weaker reactions with cells of various monkey species. How- ever, the antihuman as well as antimonkey globulin sera must be first absorbed with pooled, washed simian red cells to remove species-specific heteroagglutinins that are normally present in the sera of laboratory animals used for immunization.

Our standard procedure calls for the use of rabbits which are given injections of whole primate serum or its purified IgG fraction. Inoculum made of enriched or purified globulins is highly recommended, since the resulting antiserum has none or very little of the undesirable heteroantibodies. Separation of IgG proteins can be achieved by various methods, one of which uses Sephadex G-25 and DEAE cellulose columns. After lyophilization and reconstitution, the product is dialyzed with saline and alum-precipitated, and its pH is adjusted to 6.7–7.0. A 1.5 ml volume of globulin solution is thoroughly mixed with an equal volume of Freund's adjuvant. To obtain adequate emulsification, the mixture is stirred overnight at 4°C.

On day 0 the rabbit is injected *intramuscularly* with 0.5 ml of inoculum prepared with *complete Freund's adjuvant* at each of six sites (4 sites on the hips, 2 sites on both sides on the back of the neck).

On day 28 a boosting injection is given *subcutaneously* using 0.5 ml of inoculum prepared with *incomplete Freund's adjuvant*, at each of six sites.

TABLE 9-15. Comparison of results of titration by the saline agglutination, antiglobulin, ficinated red cell, and dextran methods

Method of titration	Titration of isoimmune rhesus monkey serum No. 216									Titration of isoimmune rhesus monkey serum No. 230									Control negatively reactive red cells
	Undil.	1:2	1:4	1:8	1:16	1:32	1:64	1:128	1:256	Undil.	1:2	1:4	1:8	1:16	1:32	1:64	1:128	1:256	
Saline agglutination	+	tr	tr	−	−	−				tr	tr	−	−	−					−
Antiglobulin	++	±	++	++	++	++	++	+±	±	++	++	++	++	++	++	+±	−		−
Ficin	+	++	++	++	++	+	tr	−	−	−	−	±	±	±	±	−	−		−
Dextran	++	++	++	+±	+±	+	−	−		+	tr	−	−	−					−

tr = trace.

When properly injected, a subcutaneous blister should appear at the site of injection.

On day 34 the animal is test-bled for 20 ml of blood without anticoagulant. Blood is drawn from an easily accessible ear vein.

The serum separated from the whole blood is then tested by titration against primate red cells coated with an appropriate type-specific univalent reagent. Unsensitized cells of the same or closely related species are used as negative controls. In most instances, one course of immunization proved sufficient to produce potent antiglobulin sera, reaching the titer of 1:500 and over, for coated cells. Nonspecific agglutination of unsensitized cells usually occurred only in the first 2–3 tubes (equivalent of 1:2 or 1:4 titers) and could be easily eliminated by twofold or threefold dilution of the antiglobulin serum.

Antiglobulin sera diluted, for economy, to 1:10 or even 1:40 are used with satisfactory results. Working dilutions should, however, be prepared prior to use and not stored for extended periods of time. In our experience, immunized rabbits sustained well weekly drawings of 30–40 ml of blood for periods of time up to 3 years, and they maintained the titer of antibodies almost unchanged from the first to the last bleedings, following only one course of immunization. In some cases, however, a gradual increase of the titer of nonspecific agglutinins was observed, so that multiple absorption with pooled unsensitized primate cells were necessary in order to prepare a reagent suitable for normal use.

Antiglobulin sera of very narrow specificity were produced by immunizing primate animals with human serum globulins. For more detailed information on the properties of antiprimate globulins and of antihuman globulin sera made in rhesus monkey, baboon, and gibbon, the readers are referred to other publications [112,196,284].

Immunization Protocol

Since the course and effects of immunization are influenced by various factors, and the procedure has to be constantly adjusted to the progress of immunization, there is no point in constructing rigid immunization schedules. Instead, protocols of a few experiments are presented here as instructive examples of actual attempts to produce isoimmune antibodies in representative species of nonhuman primates. Tables 9-16 and 9-17 illustrate the progress of immunization in a chimpanzee and a rhesus monkey, respectively. As shown in the tables, it took between 2 and 6 months and multiple injections of an antigen to produce antibodies of a titer considered adequate for blood typing purposes. But, once the sufficient titer was attained, it remained high

TABLE 9-16. Protocol of immunization of a chimpanzee

Donor: Sean, Ch-643, *Pan troglodytes*
    Male, born 1961
    Weight (at start) 39.5 kg
    Blood type: A, MN
        V.B
        cef
        G^c, H^c

Recipient: Leo, Ch-335, *Pan troglodytes*
    Male, born 1962
    Weight (at start) 46 kg
    Blood type: A, MN
        V.A
        cef
        G^c, H^c

Anticipated antibody: anti-B^c

Date	Procedure	Saline	Antiglobulin	Ficin
		Titer against donor RBCs		
24 Nov 1971	Test bleeding (preimmunization)	0	0	0
	Intramuscular injection of 1 ml of packed donor red cells mixed with 1 ml of complete Freund's adjuvant			
6 Jan 1972	Test bleeding	0	2	0
	IM injection of 1 ml of packed RBCs with 1 ml incomplete Freund's adjuvant			
3 Feb 1972	Test bleeding	2	10	4
	IM injections of 1 ml of packed RBCs with 1 ml incomplete Fruend's adjuvant			
6 Mar 1972	Test bleeding	2	12	2
	IM injection of 1 ml packed RBCs with 1 ml incomplete Freund's adjuvant			
7 Apr 1972	Test bleeding	4	12	2
	IM injection of 1 ml packed RBCs with 1 ml incomplete Freund's adjuvant			
17 May 1972	Test bleeding, plasmapheresis (600 ml)	64	512	2
20 June 1972	Test bleeding, plasmapheresis (600 ml)	32	256	1
21 July 1972	Test bleeding, plasmapheresis (600 ml)	0	32	4
18 Aug 1972	Test bleeding, plasmapheresis (600 ml)	0	16	8
12 Sept 1974	Test bleeding (preimmunization)	0	2	0
	Intraperitoneal injection of 100 ml washed, resuspended donor cells			
19 Sept 1974	Test bleeding, plasmapheresis (300 ml)	511	1,000	32
3 Oct 1974	Test bleeding, plasmapheresis (350 ml)	300	512	128
10 Oct 1974	Test bleeding, plasmapheresis (350 ml)	300	380	16
17 Oct 1974	Test bleeding, plasmapheresis (300 ml)	200	512	64
24 Oct 1974	Test bleeding, plasmapheresis (300 ml)	200	512	16
31 Oct 1974	Test bleeding, plasmapheresis (300 ml)	64	512	4
14 Nov 1974	Test bleeding, plasmapheresis (250 ml)	64	256	0
22 Nov 1974	Test bleeding, plasmapheresis (300 ml)	64	512	0
9 Dec 1974	Test bleeding, plasmapheresis (250 ml)	8	250	0
12 Dec 1974	Test bleeding, plasmapheresis (300 ml)	8	100	0
27 Dec 1974	Test bleeding, plasmapheresis (300 ml)	25	512	16
3 Jan 1975	Test bleeding, plasmapheresis (250 ml)	12	512	16
16 Jan 1975	Test bleeding, plasmapheresis (400 ml)	12	48	32
30 Jan 1975	Test bleeding, plasmapheresis (150 ml)	8	64	32

for long periods of time, thus allowing stockpiling of large quantities of antisera.

Plasmapheresis

Plasmapheresis is an experimental procedure in which blood is removed and the corpuscles separated by centrifuging, suspended in Ringer's solution, and returned to the donor's circulation while the plasma is set apart for future use. Application of this technique allows collection of impressive volumes of immune sera even from small-sized animals, without causing severe anemia by continuous reduction of the number of red cells. On the other hand, the slow depletion of the plasma proteins that may result from repeated plasmaphereses is relatively easily compensated by a healthy animal, at least for the time usually needed to complete the immunization program.

To carry out plasmapheresis, the animal is anesthetized and weighed, and a small amount of its blood is drawn to determine the hematocrit. (If it is found to be below 30, plasmapheresis is not recommended.) After the intravenous infusion of lactated Ringer's solution into the cubital vein has been accomplished, the blood is drawn from another vein (for example, the femoral vein) into a vacutainer bottle containing appropriate amounts of an anticoagulant (for example, acid citrate dextrose). To determine the maximum amount of blood in mililiters to be drawn, the weight of the animal is to be multiplied by 15. The blood is then transferred from the bottle to a plastic transfer bag (Transfer Pack by Fenwal Laboratories, Deerfield, Illinois 60015), which in turn is placed in a centrifuge and spun down at 2,000 rpm for 20 min. Separated plasma is then transferred to another transfer bag, using the plasma extractor, while the red cells are resuspended in the intravenous solution and reinfused into the animal's circulation through the previously established intravenous set.

Standardization of Typing Reagents

Whenever feasible, pairing of the recipient and donor animals in an immunization experiment is based on knowledge of their blood groups, and the specificity of antibodies to be produced can be to some extent anticipated. Not infrequently, however, the resulting antiserum proves to be a mixture of antibodies of various specificities that are to be identified and, if necessary, separated in order to prepare a reagent suitable for blood grouping purposes.

As mentioned earlier in this chapter, the simplest methods of identification of the type of antibody is to use, in parallel, various techniques of titration. To give an example, in Table 9-18 are compared the reactions of a chimpan-

TABLE 9-17. Protocol of immunization of a rhesus monkey

Donor: Cindy, Rh-928, *Macaca mulatta*   Recipient: Lois, Rh-930, *M. mulatta*
    Female, weight (at start) 4.5 kg   Female, weight (at start) 5.5 kg
    Blood type: B+F+C+D+E+G+   Blood type: B+F+C+D+E−G+
                Anticipated antibody: anti-E^{rh}

Date	Procedure	Saline	Antiglobulin	Ficin
			Titer with donor RBCs	
22 Apr 1975	Test bleeding (preimmunization)	0	0	0
	Intramuscular injection of 1 ml packed donor RBCs mixed with 1 ml of complete Freund's ajuvant			
20 May 1975	Test bleeding	0	4	1
	Intramuscular injection of 1 ml packed RBCs mixed with incomplete Freund's adjuvant			
18 June 1975	Test bleeding, plasmapheresis (80 ml)	0	128	96
25 June 1975	Test bleeding, plasmapheresis (80 ml)	1	256	256
7 July 1975	Test bleeding, plasmapheresis (80 ml)	1	64	128
24 July 1975	Test bleeding, plasmapheresis (80 ml)	0	32	128
13 Aug 1975	Test bleeding, plasmapheresis (80 ml)	0	32	64
25 Sep 1975	Test bleeding	0	16	8
	Intraperitoneal injection of 20 ml of washed, resuspended donor RBCs			
2 Oct 1975	Test bleeding, plasmapheresis (120 ml)	0	128	32
9 Oct 1975	Test bleeding, plasmapheresis (120 ml)	0	64	64
16 Oct 1975	Test bleeding, plasmapheresis (120 ml)	0	48	16
25 Oct 1975	Test bleeding, plasmapheresis (80 ml)	0	32	32
30 Oct 1975	Test bleeding	0	8	16
	Intraperitoneal injection of 20 ml of washed, resuspended donor RCBs			
7 Nov 1975	Test bleeding, plasmapheresis (120 ml)	0	96	64
14 Nov 1975	Test bleeding, plasmapheresis (100 ml)	0	96	48
21 Nov 1975	Test bleeding, plasmapheresis (120 ml)	0	64	32
29 Nov 1975	Test bleeding, plasmapheresis (120 ml)	0	32	16

zee isoimmune serum with the red cells of five chimpanzees, using three different methods of titration. No reactions were obtained by the saline agglutination method, showing that no IgM antibodies were present. Strong reactions were obtained, however, by the antiglobulin and ficin methods, indicating the presence of IgG antibodies. By the antiglobulin method, red cells of chimpanzee Walter (No. 168) reacted in very high titer, red cells from three other chimpanzees reacted in considerably lower titer, and those

TABLE 9-18. Comparison of titrations of serum of isoimmunized chimpanzee No. 4 (John) by three different methods

Test red cells of chimpanzee:			Titers by methods of:		
Number	Name	Simian-type blood group	Saline agglutination	Antiglobulin	Ficinated red cells
No. 1	Buddha	V.A. cef	0	24	1
No. 168	Walter	V.B. CeF	0	214	214
No. 64	Duane	v.B cef	0	12	0
No. 225	Andy	V.AB cef	0	32	0
No. 4	John	v.B CEf	0	0	0

of the fifth failed to react. On the other hand, titrations by the ficinated red cell method gave clear reactions and high titers only with the red cells of chimpanzee No. 168. Thus, the serum tested contained antibodies of at least two distinct specificities, one of a very high titer for the factor F^c of the chimpanzee R-C-E-F system, and the second weaker and reacting only by antiglobulin and not by the ficinated red cell method, and, therefore, presumably detecting a specificity belonging to the V-A-B-D chimpanzee blood system. Of course, the antibodies in this serum might have proved separable by absorption with appropriate chimpanzee red cells, but by the use of more than one method of titration necessary information could be obtained with less effort.

An instructive example of the technique of analysis of unknown antiserum by use of various testing methods, combined with absorption and fractionation experiments, was described in an earlier paper [126]. In that case, antiserum from an isoimmunized chimpanzee, Leo, was titrated by saline, antiglobulin, and ficin techniques, at $37°$ as well as $4°C$, with a series of chimpanzee red cells of various known blood types. The results of the tests, shown in Table 9-19 were interpreted as follows:

1. The serum of Leo reacted at body temperature, both by the saline and by antiglobulin methods, in highest titers, with red cells of the B^c-positive type, thus indicating that isoimmune serum contained antibodies of anti-B^c specificity, as anticipated from the fact that chimpanzee Leo was of type V.A, cef, G,H,L while the donor, Sean, was of type V.B,cef,G,H,l.

2. In addition to B^c-positive red cells, also cells bearing D^c agglutinogen reacted with the Leo's serum although to a somewhat lower titer. Red cells lacking both B^c and D^c remain unagglutinated. This was interpreted to mean that Leo's serum contained not only anti-B^c but also anti-Y^c, an antibody that is cross-reactive for all chimpanzee red cells having agglutinogens B^c or D^c

TABLE 9-19. Results of titration of the serum of an isoimmunized chimpanzee (Leo, No. 335); and the presumed specificities of the antibodies detected

| Tested against red cells of chimpanzee: | | Simian-type blood groups | | | | | | | | Titers against: | | | | | |
| | | | | | | | | | | Unmodified red cells | | | Ficinated red cells | | |
Name	No.	V-A-B	C-E-F	Gc	Hc	Kc	Lc	Nc		At 4°C	At 37°C	Anti-globulin	At 4°C	At 37°C	Anti-globulin
Leo	335	V.A	cef	+	+	+	+	−		128	3	0	48	1	10
Sean (donor)	643	V.B	cef	+	+	+	−	−		192	200	512	48	128	40
Amos	9	v.AB	CEF	+	+	+	+	+		100	200	512	48	128	40
Tom	11	V.A	cef	+	−	tr	−	−		64	2	2	48	128	11
Edgar	13	V.O	cef	+	+	+	+	−		80	1	0	32	32	16
Gabriel	19	v.AD	cef	+	+	+	+	−		100	128	96	56	48	22
Man															
Adult group O										1½	0	nd	3	1	nd
Cord group O										1½	0	nd	6	1	nd
Presumed antibody specificities										Cold auto-antibody	Anti-Bc, anti-Yc	Anti-Bc, anti-yc	Cold auto-antibody	Anti-Oc, anti-Pc	Auto-antibody (nonspecific)

nd = not done; tr = trace.

or both. It was possible to confirm the validity of that conclusion by absorption experiments carried out with D^c-positive B^c-negative cells that resulted in production of monospecific anti-B^c reagent (Table 9-20).

3. The serum of Leo tested at 37°C against ficinated red cells reacted strongly with the red cells of some but not all chimpanzees, but the specificity of the reactions was different from that against unmodified red cells. Apparently, the antibody demonstrated by the ficinated red cell method is of a specificity not only unrelated to the V-A-B system, but also unrelated to the C-E-F system and the other simian-type blood groups listed in Table 9-20. Therefore, it was assigned the new symbol anti-O^c.

4. When the pure anti-B^c reagent (Leo's serum absorbed with D^c-positive red cells of chimpanzee Gabriel) was tested at 37°C against ficinated red cells, it no longer reacted with modified red cells of Gabriel, though it continued to react in moderately high titer with the red cells of other chimpanzees. This indicated that the anti-O^c in Leo's serum could be fractionated by absorption, leaving behind a separate antibody, tentatively designated anti-P^c, that defined still another specificity, demonstrated on ficinated chimpanzee red cells but not unmodified chimpanzee red cells. This observation might be compared to the fractionation of human anti-A serum by absorption with human A_2 cells.

5. Reagents of newly defined specificities anti-O^c and anti-P^c were tested with the modified red cells of 45 chimpanzees (Table 9-21) and 39 of them (86.7%) proved to have both specificities, O^c and P^c, while 32 (71.1%) were shown to have specificity P^c alone. Inspection of the data in Table 9-21 led to the conclusion that "the serum anti-P^c defined a new blood group system, independent from all other simian-type blood group systems so far known in chimpanzees (V-A-B, C-E-F, G^c, H^c, L^c, N^c)."

The last statement was inaccurate since, a few years later, analysis of the distribution of blood groups in a sample of over 200 chimpanzees established a strong association of P^c with the specificities of the chimpanzee R-C-E-F blood group system [202]. Table 9-22 reproduces the 2×2 contingency tests that showed that P^c was negatively associated with L^c and C^c, but that a highly significant positive association existed between P^c and c^c. Since reactions with anti-P^c were always negative when c^c was absent but c^c was sometimes present without P^c, it was assumed that the relationship between P^c and c^c was a graded one, comparable to the relationship between A_2 and A_1 in the A-B-O blood group system. This assumption was confirmed by subsequent absorption experiments, thus justifying inclusion of P^c in the R-C-E-F system, as the subgroup of c^c. Accordingly, the name anti-P^c was

TABLE 9-20. Titration of serum of chimpanzee Leo, No. 335, before and after multiple absorptions with the red cells of D^c-negative B^c-negative type of chimpanzee Gabriel, No. 19 (composite table of findings)

Tested against red cells of chimpanzee:		Simian-type blood groups	Serum before absorption tested by method of:			Serum after absorption tested by method of:		
Name	No.		Saline agglutination	Antiglobulin	Ficinated red cell	Saline agglutination	Antiglobulin	Ficinated red cell
Pan troglodytes								
1. Leo	335	V.A,cef,G,H,L	3	0	1	0	0	0
2. Sean (donor)	643	V.B,cef,G,H,l	200	512	128	nd	nd	nd
3. Gabriel	19	v.AD,cef, G,H	128	96	48	0	0	0
4. Amos	9	v.AB,Ccef,G,H	200	512	0	32	350	0
5. Edgar	13	V.O,cef,G,h	1	0	32	0	½	40
6. Ellis	15	V.A,Ccef,g,H	nd	nd	nd	0	0	24
7. Mack	17	V.B,cef,G,H	nd	nd	nd	32	500	14
8. Lindsey	192	v.AB,cef,G,H	48	512	32	32	500	32
9. Herbie	191	v.BD, cef,G,H,l	250	500	24	32	500	24
10. Cassius	209	V.B,CCEF,G,H,L	nd	nd	nd	32	500	0
11. Andy	225	v.AB,cef,G,h	200	500	32	32	500	10
12. Tom	11	V.A,cef,G,H,l	2	2	128	nd	nd	nd
Pan paniscus								
13. Kitty		v.D,CCEf,g,H	nd	nd	nd	0	0	0
Presumed specificities			Anti-B^c + Y^c and autoantibody	nd	Anti-O^c + anti-P^c	Anti-B^c	Anti-B^c	Anti-P^c

nd = not done.

TABLE 9–21. Results of test with anti-O^c+P^c and anti-P^c reagents carried out by enzyme-treated red cell method and blood samples of 45 chimpanzees

Name	No.	Human type	Simian type	Anti-O^c + anti-P^c	Anti-P^c
1. Leo	335	A	V.A,cef,G,H,L	−	−
2. Sean	643	A	V.B,cef, G,H,K,n	+	+
3. Gabriel	19	A	v.AD,cef,G,H,L	+	−
4. Blackie	A-30	A	V.O,cef,G,H,l	+	+
5. Mary	A-86	A	V.D, CEF,g,H	+	+
6.	A-276	A	v.A,CEF,G,H,L	−	−
7. Perry	A-27	A	v.A,cef,g,H,L	+	+
8. Samantha	A-28	A	v.B,cef,G,H,L	+	+
9. Ken	A-75	A	v.BD,cef,G,H,L	+	+
10. Batsy	A-76	A	V.A,CcEf,g,H,l,n	+	+
11.	A-274	A	V.B,cef,g,h,l	+	+
12.	A-275	A	v.AB,cef, G,H,l	+	+
13.	A-277	A	v.BD,cef,G,h,l	+	+
14. Amos	9	A	v.AB,CcEF,G,H,L,n	−	−
15. Ellis	15	A	V.A,CcEf,g,H,l,n	+	+
16. Edgar	13	A	V.O,cef,G,h,L,n	+	+
17. Pudding	170	A	v.AB,cef,G,h,L,n	−	−
18. Moses	171	A	V.B,cef,G,H,l,n	+	+
19. Ardein	172	A	V.A,CCeF,G,h,L,n	−	−
20. Hala	174	A	v.A,CceF,G,h,l,n	+	+
21. Ursula	176	A	v.A,CcEf,G,H,L,n	+	+
22. Mabel	178	A	v.AB,cef,g,h,l,n	+	+
23. Jamel	175	A	V.A,cef,G,H,L,n	+	−
24. Rufe	114	A	v.B,cef,g,H,l,n	+	−
25. Lindsay	192	A	v.AB,cef,G,h,L	+	+
26. Oscar	211	A	V.A,CcEf,G,H,L,n	+	+
27. Paul	277	A	v.AB,ced,G,H,L,n	+	−
28. Bob	179	A	v.B,cef,G,h,L,n	+	−
29. Max	181	A	v.B,cef,G,h,l,n	+	−
30. Roger	183	A	v.B,cef,G,H,l,n	+	+
31. Rusty	185	A	v.A,cef,G,h,L,n	+	+
32. Ahab	189	A	v.BD,cef,G,h,L,n	+	+
33. Kurt	191	O	v.D,CceF,G,H,l,n	+	+
34. Wanda	196	A	V.A,cef,G,h,l,n	+	+
35. Panda	110	A	V.A,cef,G,H,l,n	+	+
36. Elsa	112	A	V.B,cef,G,h,L,n	+	+
37. Edwinna	118	A	v.A,cef,G,h,L,n	+	+
38. Lolita	120	A	V.O,cef,G,H,L,n	+	+
39. Joey	123	A	V.D,cef,g,H,l,n	+	+
40. Andrea	124	A	v.BD,CCEf,G,h,L,n	+	−
41. Elmer	127	A	v.AB,cef,g,h,l,n	+	+
42. Damon	129	A	V.A,cef,G,h,L,n	+	+
43. Archie	121	A	v.A,cef,G,h,l,n	+	+
44. Abbey	206	O	v.B,CCeF,G,H,L,N	−	−
45. Tom	11	A	V.A,cef,G,H,l,n	+	+

changed to anti-c_1^c, and the reagent thus defined became a part of the panel of standard R-C-E-F antisera used for routine typing of chimpanzee blood.

The above chronology of the development of a typing reagent obtained from the serum of an isoimmunized chimpanzee is offered as an illustration of the chain of procedures, serological and statistical, that lead, step by step, to the definition of a new specificity and of its position among other, already established blood groups of a given species. For details of the application of statistical methods for analysis of serological data, the reader is referred to other publications [44,83].

As mentioned earlier in this chapter, the parallel use of various testing techniques allows the detection of various antibodies in the immune sera, thus enlarging the scope of applications of an antiserum. On the other hand, the red cell specificities also can be classified on the basis of a testing technique that brings about agglutination of the red cells bearing the given antigen. For example, those specificities that are reactive by the antiglobulin method but not by the ficinated red cell technique are presumed to belong to systems analogous to the human M-N-S blood group system. Specificities that are reactive by the ficin as well as the antiglobulin technique presumably belong to a blood group system analogous to the human Rh-Hr system, etc.

One pitfall is the occurrence of a prozone in titrations by the enzyme-treated red cell method of freshly obtained rhesus monkey sera that is not observed by us in other primate species. Obviously, if only one-tube tests were done with such antisera, false-negative reactions could result. If the serum is inactivated, however, by heating for half an hour at 56°C, the

TABLE 9-22. Contingency tests for associations among specificities $L^c(R^c)$, C^c, c^c, and P^c (c_1^c)

	C^c			c^c			P^c (c_1^c)			$\chi^2_{(1)}$
	+	−	Total	+	−	Total	+	−	Total[a]	
$L^c(R^c)$ +	94	91	185	167	18	185	65	57	122	$L^c:C^c$ $\chi^2 = 13.7959$
										$P < 0.001$
$L^c(R^c)$ −	28	72	100	97	3	100	71	10	81	$L^c:c^c$ $\chi^2 = 4.3071$
										$0.05 > P > 0.02$
C^c +				101	21	122	46	48	94	$C^c:c^c$ $\chi^2 = 30.2892$
										$P < 0.001$
C^c −				163	0	163	90	19	109	$L^c:P^c$ $\chi^2 = 26.015$
										$P < 0.001$
c^c +							136	45	181	$C^c:P^c$ $\chi^2 = 39.017$
										$P < 0.001$
c^c −							0	22	22	$c^c:P^c$ $\chi^2 = 50.0846$
										$P < 0.001$

[a]Only 203 blood samples tested with anti-P^c serum.

prozone disappears (Table 9-23), showing that it was due to interference from complement present in the fresh serum and not to so-called "optimal-proportion." The presence of complement is sometimes apparent when hemolysis occurs in the first few tubes of the titration. Instead of inactivating the serum, the titration can be converted to the antiglobulin method, by washing the red cells in each tube and then adding antiglobulin serum, whereupon the prozone disappears. Evidently, there are no prozones when the antiglobulin method is used, whether the red cells are ficinated or tested untreated.

Isoimmune sera can be used not only for blood typing within the same species, but also in closely related species; for example, isoimmune rhesus sera can be used for typing red cells of stump-tailed macaques (*Macaca arctoides*), crab-eating macaques (*M. fascicularis*), pig-tailed macaques (*M. nemestrina*), several other species of macaques [124], and even baboons [125]. Baboon isoimmune sera, on the other hand, can be used for typing red cells of macaques [208]. It should be noted that two isoantisera that show parallel polymorphism within their own species can differ in their reactions with red cells from animals of another species. For example, in Table 9-24 are shown reactions of two anti-A^P isoimmune baboon sera tested against red cells from *Macaca mulatta, M. fascicularis,* and *Theropithecus gelada.* It is obvious that the two antisera differ in their specificity even though they gave parallel reactions in tests on red cells from all baboons so far examined.

As another example, Table 9-25 shows three pairs of rhesus isoimmune antisera that gave strictly parallel reactions when tested with rhesus monkeys but discordant reactions in tests with red cells of some other species of macaques. This phenomenon of "split specificity" of the cross-reactive sera illustrates the vagueness of the notion of monospecificity. It also demonstrates that the so-called *serologic criterion of unity* advocated by some authors [30,38,68,153,219] may be actually unachievable. The serological criterion of unity is presumed to be satisfied when erythrocytes from positively reacting monkeys absorb all antibody from an antiserum. Yet, as shown in our experiments, there is always a possibility that positively reacting cells from related species can still be found that will absorb only part of the antibodies, leaving behind a fraction still reactive with another positively reacting cell.

A different kind of reagent is produced by immunization with blood from closely related primate species. For example, chimpanzee cross-immunized with human red cells produced antibodies that not only were reactive with human red cells, but also detected individual differences on chimpanzee erythrocytes. Antisera produced by immunization of baboons with the red cells of geladas were found to be type-specific in tests with baboon red cells, etc. Since the use of these cross-immune antisera does not require separate techniques of testing, they will not be further discussed here.

TABLE 9-23. Elimination of prozone by inactivation of isoimmune rhesus monkey serum No. 228

Method of titration	Condition of serum	Undil.	1/2	1/4	1/8	1/16	1/32	1/64	1/128	1/256	Negative control red cells
Saline agglutination	Fresh	−	−	−							−
Antiglobulin	Fresh	+++±	+++±	+++±	+++±	+++±	++	++	+±	tr	−
Ficin	Fresh	−	−	−	tr	+	+±	+	tr	tr	−
Ficin	Inactivated	+++±	++	++	++	++	+±	tr	−	−	−

tr = trace.

TABLE 9-24. Cross-reactions of two anti-A^P isoimmune baboon sera, with red cells of macaques and gelada monkeys (saline agglutination method)

Antiserum	Papio cynocephalus, animal numbers:				Macaca mulatta, animal numbers:				Macaca fascicularis, animal numbers:				Theropithecus gelada,[a] animal number:	
	2	9	11	13	208	222	228	230	60	160	168	98	5	6
Anti-A^P (B-2)	−	++	++	−	+++	++	+++±	+++±	++	−	++	−	−	−
Anti-A^P (B-11)	−	++	++	−	+++±	−	++±	−	−	−	−	−	−	−

[a]While geladas gave negative reactions for A^P, they were both positive with other baboon isoimmune sera.

TABLE 9-25. Comparison of the distribution of red cell specificities, defined by pairs of rhesus isoimmune sera of supposedly identical specificities, in six species of macaques

Antiserum		Frequency (%) of positive reactions with red cells of:					
Identification	Specificity	*M. mulatta*	*M. nemestrina*	*M. fascicularis*	*M. arctoides*	*M. radiata*	*M. sylvanus*
Rh-2	Anti-B^{rh}	90.29	4.5	70.2	0.0	100.0	100.0
Rh-216	Anti-B^{rh}	90.29	96.6	53.2	100.0	100.0	100.0
Rh-212	Anti-D$_2^{rh}$	34.41	39.3	0.0	0.0	0.0	0.0
Rh-924	Anti-D$_2^{rh}$	34.41	39.3	0.0	30.0	n.t.	87.5
Rh-321	Anti-N^{rh}	96.76	100.0	100.0	100.0	n.t.	100.0
Rh-932	Anti-N^{rh}	96.76	100.0	21.4	100.0	n.t.	100.0

n.t. = not tested.

Typing Reagents

As the result of immunization, comparative titrations, fractionation through absorption, and multiple testing with large panels of red cells obtained from animals of the same or closely related species, panels of typing reagents of various specificities are developed for routine use in blood grouping of apes and monkeys. For each specificity, panels of positively and negatively reacting control cells are established that are to be included with each series of newly tested blood specimens of a given primate species.

Table 9-26 lists rhesus isoimmune sera currently employed by us for blood grouping macaques of various species, namely rhesus monkeys (*Macaca mulatta*) [124,211,287], pig-tailed macaques (*M. nemestrina*) [204], crab-eating macaques (*M. fascicularis*) [142], stump-tailed macaques (*M. arctoides*) [125], bonnet macaques (*M. radiata*) [205], and Barbary macaques (*M. sylvanus*) [198].

As mentioned earlier in this book, there were other successful attempts to produce rhesus blood typing reagents by immunization with monkey red cells. One such program was carried out by Edwards [38-41], who established a panel of hemagglutinating sera of isoimmune or cross-immune and heteroimmune origins. The reagents detected 16 different specificities which, on the basis of population as well as family studies, were classified in nine blood group systems: the G and H rhesus blood group systems [38], the I system [39], and the rhesus monkey blood group systems J, K, L, M, N, and O [40]. Using the same pairs of monkeys for immunization, Duggleby and Stone [30,31] produced another set of rhesus isoimmune typing reagents for which they proposed their own nomenclature, even though specifities of some of their sera were identical with those produced earlier by Edwards, while the six blood group systems they defined (H,I,G,J,L, and K) somewhat

TABLE 9-26. Rhesus isoimmune sera used for blood
grouping tests in macaques

No.	Immunized animal		Specificity of the reagent
1.	Rh-1	Caesar	A^{rh}
2.	Rh-2	Eve	B^{rh}
3.	Rh-216	Path	B^{rh}
4.	Rh-4	Eleanor	C^{rh}
5.	Rh-6	Antigone	D^{rh}
6.	Rh-942	Kate	D_3^{rh}
7.	Rh-212	Mino	D_2^{rh}
8.	Rh-924	Fanny	D_2^{rh}
9.	Rh-218	Minim	D_1^{rh}
10.	Rh-224	Temple	D_1^{rh}
11.	Rh-228	Harper	$D_1^{rh} + X$
12.	Rh-930	Lois	$D_1^{rh} + Y$
13.	Rh-3	Barney	F^{rh}
14.	Rh-178	Rachel	G^{rh}
15.	Rh-926	Sally[a]	$J_3^{rh} + v$
16.	Rh-926	Sally[b]	J_3^{rh}
17.	Rh-1062	Avery	J_2^{rh}
18.	Rh-796	Osmia[a]	$J_1^{rh} + Z$
19.	Rh-796	Osmia[c]	J_1^{rh}
20.	Rh-938	Camille[d]	J^{rh}
21.	Rh-946	Laura	L^{rh}
22.	Rh-936	Dana	M^{rh}
23.	Rh-321	Harry	N^{rh}
24.	Rh-932	April	O^{rh}
25.	Rh-922	Jane	P^{rh}
26.	Rh-928	Cindy	S^{rh}

[a]Unabsorbed.
[b]Absorbed with RBCs of Rh-928 or Rh-934.
[c]Absorbed with RBCs of Rh-934 or Rh-936.
[d]Absorbed with RBCs of Rh-796 or Rh-2.

overlapped those described by Edwards [41,220]. No attempts were made to reconcile the two sets of notations and it has not been possible up to now to carry out satisfactory comparisons of specificities detected by reagents of these authors and the specificities of our own rhesus immune sera. Edwards' sera were not available for comparative testing. The reagents used by Duggleby and Stone give clear reactions in their own hands, but we found them difficult to work with because of their low titer (1–2 units) and low avidity. Moreover, the reagents were available to us in quantities too small to make extensive comparative testing possible. However, limited comparative studies of rhesus isoimmune sera of various sources were carried out by Duggleby

and Stone [31] and by Sullivan et al. [222]. Significant positive correlation
was claimed between their anti-K-1 reagents and two of the sera produced by
Hirose and Balner [64] despite the fact that the two sets of reagents were
tested by different techniques, namely saline agglutination and the dextran/
BSA method, respectively. On the other hand, the anti-D^{rh} serum first
described by Moor-Janowski et al. [134] supposedly contained fractions of
anti-K-1 and anti-H-2 specificities, as defined by Duggleby's reagents. Antisera of several new specificities were described more recently by Sullivan et
al. [224,225], but it is not known whether any comparative testings with the
rhesus reagents of other origins were attempted.

Table 9-27 lists the reagents currently used by us for typing baboon red
cells. Most of the antisera are of isoimmune origin, but a few were obtained
from the sera of baboons immunized with the red cells of other species,
namely with those of a gelada [138] and those of a human donor (unpublished). Interestingly enough, those cross-immune baboon sera closely duplicated the specificities of the isoimmune sera. Even more interesting is the
fact that one of our rhesus isoimmune sera that cross-reacted with baboon
red cells was found to define a specificity that is part of the so-called baboon
B^p graded blood group system (see p. 94). It is noteworthy that some of the
spontaneous antibodies occurring in "normal" baboon sera share specificities
with some of our antisera that were obtained by intentional immunization of
baboons. All those facts suggest that antigens of baboon red cells also occur,
in identical or very similar form, on red cells of other species of primates,
including man. Identical or similar antigenic structures may also constitute a
part of other ubiquitously present substances to which baboons are acciden-
tally exposed.

Among the baboon antisera listed in Table 9-27, particular importance was
ascribed to the reagents that defined the so-called "ca" and "hu" specificities.
The anti-ca antisera produced by immunization of chacma and hamadryas
baboons with the red cells of yellow baboons were supposed to be specifically
directed against antigenic structures present on the red cells of all yellow
(*Papio cynocephalus*) and olive (*P. anubis*) but not on the red cells of
hamadryas (*P. hamadryas*) or chacma (*P. ursinus*) baboons. On the other
hand, anti-hu produced by injecting red cells of a chacma baboon into an
olive baboon was supposed to be a species-specific reagent selectively reacting with red cells of chacma and hamadryas baboons, but not with those of
olive or yellow baboons [139]. In fact, blood group "ca," originally believed
to be species-specific, was later found to be polymorphic in both olive and
hamadryas baboons, as well as in the hybrids; "hu," on the other hand,

TABLE 9-27. Baboon blood / grouping reagents

No.	Immunized animal	Donor	Specificity
1.	B-2 Paula, olive baboon	B-11 Chubby olive baboon	A^P
2.	B-13 Jack, yellow baboon	B-9 Eugene, yellow baboon	$A^P + Y^P$
3.	B-127 Blue, hybrid	B-15 Irwin, yellow baboon	$A^P + X^P$
4.	B-9 Eugene, yellow baboon	B-13 Jack, yellow baboon	$B^P + Z^P$
5.	B-9 Eugene[a]	B-13 Jack	B_1^P
6.	B-15 Irwin, yellow baboon	B-127 Blue, hybrid	$B_2^P + O^P$
7.	B-15 Irwin[b]	B-127 Blue	B_2^P
8.	Rh-212 Mino,[c] rhesus monkey	Rh-214 Iodine, rhesus monkey	$B_3^P (D_3^{rh})$
9.	B-77 Chatta, olive baboon	G-5 Benton, gelada	$B^P + G^P$
10.	B-11 Chubby, olive baboon	B-2 Paula, olive baboon	$C^P + Q^P$
11.	B-11 Chubby[d]	B-2 Paula	C^P
12.	B-32 Lynn, olive baboon	B-60012 Janna, olive baboon	G^P
13.	B-268 Grayma, olive baboon	B-85 Nosmo, chacma baboon	hu
14.	B-176 Camilla, hamadryas baboon	B-252 Eberle, yellow baboon	ca
15.	B-87 Souds, chacma baboon	b-553 Benita, yellow baboon	ca + E^P
16.	B-60012 Janna, olive baboon	B-32 Lynn, olive baboon	N^P
17.	B-17 Janny, yellow baboon	B-133 Roman, hybrid	P^P
18.	B-131 Ralph, hybrid	B-776 Ilex, hamadryas baboon	$S^P + T^P$
19.	B-131 Ralph[e]	B-776 Ilex	S^P
20.	B-133 Roman, hybrid	B-17 Janny, yellow baboon	U^P
21.	B-143 Dave, hybrid	B-230 Vivian, yellow baboon	$V^P + V_1^P$
22.	B-143 Dave[f]	B-230 Vivian	V^P
23.	B-145 Thorb, hybrid	B-774 Yucca, hamadryas baboon	L^P
24.	B-776 Ilex,[g] hamadryas baboon	B-131 Ralph, hybrid	M^P
25.	B-808 Agnes,[h] hybrid	Human O, Rh pos.	B_2^P

[a]Absorbed with pooled blood of B-2 and B-11.
[b]Absorbed with blood of B-2.
[c]Cross-reacting rhesus isoimmune serum.
[d]Absorbed with pooled blood of B-9 and B-13.
[e]Absorbed with blood of B-9.
[f]Absorbed with pooled blood of V_2 type.
[g]Absorbed with blood of B-790.
[h]Transfused with human blood.

which was believed to be a species characteristic of hamadryas and ursinus baboons, was also detected in some but not all olive baboons [209].

A set of reagents routinely used for blood grouping tests in chimpanzees, and also to some extent in other anthropoid apes, is presented in Table 9-28. As can be seen, groups of reagents of the same specificity but of various origins are employed in parallel to serve as an internal control of tests, and also to reveal individual differences among red cells that seem to be of identical specificities, as occurs not too infrequently in tests with primate red cells. Such differences, if present, point to the mosaic nature of antigens previously believed to be indivisible.

Limited isoimmunization experiments carried out in gibbons resulted in production of blood grouping reagents designated anti-A^g, anti-B^g [133], and anti-C^g [271]. The two former reagents react by all three serological methods, i.e., the saline agglutination, the antiglobulin, and the ficinated red cell methods. Anti-A^g was produced by two gibbons of the subspecies *Hylobates lar pileatus* that had been injected with blood of the subspecies *Hylobated lar lar*; the factor A^g appears to be characteristic for all animals of the subspecies *H. lar lar*. On the other hand, factor B^g appears to be characteristic of the subspecies *H. lar pileatus*. The blood factor C^g belongs to a different category. The anti-C^g sera, which were produced in two *H. lar lar* gibbons, define individual differences within their species. Anti-C^g reacts in high titers by the antiglobulin technique and hardly at all by the saline or ficin methods. Experiments carried out so far with the available gibbon antisera concerned only very small numbers of animals, and no association studies were possible to ascertain relationships among the three specificities.

By immunization of rabbits with red cells of Celebes black apes (*Cynopithecus niger*), an antiserum was obtained that, after absorption to remove species-specific heteroagglutinins, defined individual differences in the red cells of that species [136]. The blood factor detected by the reagent has been designated A^{ba}, where ba stands for "black ape," and the corresponding antibody is designated anti-A^{ba}. The relative ease with which the anti-A^{ba} serum was produced in rabbits is comparable to the original experiments of Landsteiner and Levine in which they discovered the human MN types [90] and of Owen and Anderson's experiments with rhesus monkey blood [156]. The main difficulty of the method is that the initial experiments must be done blind, in that not only the blood used for the immunization, but also that used for absorptions has to be selected at random. Therefore, whenever possible, the method of iso- or cross-immunization should be the technique of choice for producing antiprimate blood typing reagents.

TABLE 9-28. Standard reagents used for chimpanzee blood grouping

No.	Reagent specificity	Origin
1.	A	Human hyperimmune
2.	B	Human hyperimmune
3.	A_1	Human hyperimmune
4.	A_1	Lectin (*Dolichos biflorus*)
5.	H	Lectin (*Ulex europaeus*)
6.	M	Rabbit heteroimmune (baboon)
7	M	Rabbit heteroimmune (man)
8.	N	Rabbit heteroimmune (man)
9.	N^V	Lectin (*Vicia unijuga*)
10.	N^V (V^c)	Lectin (*Vicia graminea*)
11.	V^c	Chimpanzee cross-immune (man) (Ch-169 Possum)
12.	$V^c + W^c$	Chimpanzee isoimmune (Ch-639 Dina)
13.	V_1^c	Chimpanzee isoimmune (Ch-19 Gabriel)
14.	A^c	Chimpanzee isoimmune (Ch-194 Herbie)
15.	B^c	Chimpanzee isoimmune (Ch-355 Leo) (absorbed)
16.	D^c	Chimpanzee isoimmune (Ch-192 Lindsay)
17.	$B^c + D^c$	Chimpanzee isoimmune (Ch-355 Leo) (absorbed)
18.	$A^c + B^c + D^c$	Chimpanzee isoimmune (Ch-336 Stu)
19.	M^c	Chimpanzee cross-immune (gibbon) (Ch-208 Mandy)
20.	R^c	Chimpanzee isoimmune (Ch-643 Sean), ETC[a]
21.	R^c	Chimpanzee isoimmune (Ch-177 Karen), ETC[a]
22.	R^c	Chimpanzee isoimmune (Ch-11 Tom) Absorbed, ETC[a]
23.	$R^c + G^c$	Chimpanzee isoimmune (Ch-490 Chica) ETC[a]
24.	C^c	Chimpanzee isoimmune (Ch-491 Bonnie), ETC [a]
25.	E^c	Chimpanzee isoimmune (Ch-38 Doug) ETC[a]
26.	E^c	Chimpanzee isoimmune (Ch-11 Tom), absorbed, ETC[a]
27.	F^c	Chimpanzee isoimmune (Ch-34 Jack), ETC[a]
28.	F^c	Chimpanzee isoimmune (Ch-136 Hope), ETC[a]
29.	$C^c + F^c$	Chimpanzee isoimmune (Ch-225 Andy)
30.	c^c	Chimpanzee isoimmune (Ch-114 Rufe), ETC[a]
31.	c_1^c	Chimpanzee isoimmune (Ch-355 Leo) absorbed, ETC[a]
32.	$O^c + c_1^c$	Chimpanzee isoimmune (Ch-335 Leo) ETC[a]
33.	H^c	Chimpanzee isoimmune (Ch-488 Ginger), ETC[a]
34.	K^c	Chimpanzee cross-immune (man) (Ch-85 Billy), ETC[a]
35.	T^c	Chimpanzee isoimmune (Ch-17 Mack), ETC[a]

[a]ETC = used by enzyme-treated red cell technique.

MONOCLONAL ANTIBODIES

The advent of monoclonal antibodies has opened exciting perspectives for potentially unlimited sources of antibodies of very narrow and precise specificities. As shown by successful attempts to produce monoclonal anti-A

[118,234] and anti-B [188] antisera, hybridoma cells can become in the near future a source of potent human blood grouping reagents not only for special research purposes but also with many commercial applications. The question arises whether this ingenious and promising method can also be applied to the production of agglutinating antisera for tests with the red cells of nonhuman primates.

The existing techniques for obtaining monoclonal antibodies are derived, by and large, from the methodology originally described by Kohler and Milstein in 1975 [77], in which antibody-forming hybrids are obtained by the fusion of the spleen cells of an immunized mouse with cultured mouse myeloma cells. (For an up-dated review of the state of art in the production of monoclonal antibodies the reader is referred to the recent article by Diamond et al. [25]; for an in-depth study of the subject we recommend the article by Milstein [121] or the monograph edited by Kennet et al. [76]). If the goal is to produce antibody against a particular red cell specificity, a mouse will be immunized with the red cells carrying that specificity. The mouse will make antibodies against the entire spectrum of red cell membrane antigens, and if the fusions of the recipient's spleen cells with myeloma cells are successful, one can expect to obtain hybridoma cells, producing antibodies of a variety of specificities. If red cells from the same donor are used as target cells in screening the hybridomas, many positive hybrids will be identified, but few if any will react specifically with the desired antigen molecule. Selection and isolation of the desired hybridoma cells for cloning will therefore require large panels of target cells of various types, ones that contain the desired specificity and others that are as similar as possible but lack the targeted specificity. Only hybrids producing antibodies that react strongly with the positive target, and not at all with the negative target, are of the desired activity. In view of the multiplicity of antigens detectable on the red cells of apes and monkeys, the limited availability of adequate panels of target cells may constitute a serious difficulty in the production of monoclonal blood grouping reagents of specificities other than anti-A or anti-B. Unfortunately, this obstacle cannot as yet be overcome by the use of purified simian-type antigens as immunogens in the way pure A, B, or H group substances can be used as immunogens in the course of the production of monoclonal A-B-O reagents.

Another possible difficulty in the production of monoclonal antibodies against simian red cell antigens may result from relatively low immunogenicity of some of the simian red cell antigens. Weaker immunogens usually stimulate lesser numbers of antibody-forming spleen cells, thus diminishing the yield of antibody-producing hybridomas that are type-specific. A primate-to-mouse immunization will probably result in the situation when a great majority of antibody-forming spleen cells produce heterospecific antiprimate antibodies while only a few cells are found producing the desired type-specific antibodies.

Very high titers of monoclonal antibodies may, surprisingly, present some conceptual difficulties: such antibodies can sometimes reveal previously unrecognized cross-reactivity. For example, they may react with previously negative donor cells, making it difficult to compare the specificity of the monoclonal antibody with that of reference antiserum.

Conventional antiserum may contain antibodies against contaminating antigens or against several antigenic determinants on a single macromolecule. Both types of antibodies can often be absorbed out of antisera to provide monospecific reagents. Monoclonal antibodies will be directed to a single antigenic site, and cross-reactions will result from one antibody's recognizing related antigenic determinants on different macromolecules. The titer of a monoclonal against a cross-reacting antigen could be higher than the titer against the immunizing antigen. Cross-reactivity cannot be removed from monoclonal antibodies by absorption.

Since, in many cases, homogenous antibodies are arduous to obtain, and since their use often presents certain problems, it is reasonable to ask whether it is really worth the effort. In the situation of blood grouping antisera for use on red cells of nonhuman primates where the tests are carried out most often on relatively small numbers of animals of a given species, the usefulness of the conventional antisera can be stretched, thanks to their cross-reactivity with the red cells of closely related species. This is, for instance, the case of the rhesus immune sera of the D^{rh} specificities that proved suitable for tests in at least six different species of macaques. The monoclonal antibodies, on the other hand, will most probably react only with antigen-bearing homologous cells of the species against which the antibody was originally produced. Bearing in mind the fact that very large numbers of hybridomas will have to be produced in order to find a few that will form the desired type-specific antibodies, the usefulness of the desired monoclonal type-specific antibodies, if they are obtained, will be limited to relatively small numbers of blood samples originating from one species of animals, and the unit price of tests with such monoclonal antibodies may prove excessively high, actually too prohibitive to justify their use for routine blood grouping tests in primate animals. This does not exclude the possibility that in the future, with the perfection of techniques of monoclonal antibody production, this kind of reagent may also become a generally accepted tool for typing apes and monkeys. Progress in this field seems to depend largely on improvement of the fusion techniques that will make it possible to obtain antibody-producing hybridomas from human or primate myeloma cells and primate lymphocytes or plasma cells.

Although the possibility of routine use of monoclonal antibodies for primate blood grouping still seems remote and not of the highest urgency, there are reports indicating that cross-reactive monoclonal reagents specifically developed to test human lymphocyte antigens can be successfully applied to the study of phylogenetic relationships among primates. A recent study by Haynes et al. [63a] showed that monoclonal antibodies to human T lymphocytes displayed different levels of cross-reactivity with determinants of T-cell subset antigens of various species of anthropoid apes and Old and New World monkeys. The reactivity of gorilla and chimpanzee T cells with the panel of reagents was identical to that of humans. Orangutan cells were reactive with 10 of 11 reagents, and gibbon cells reacted with 9 of the 11 reagents used. In contrast, cells of all Old World monkey species tested were similar and reacted with only 5 of the 11 human T-cell-specific reagents. Cells from New World monkey species reacted only with antibody to the inducer T cell. Those results could be conceived, according to the authors, as additional evidence for a relatively recent common ancestor for man, chimpanzee, and gorilla, and as additional support for the existing proposed ranking of evolutionary divergence of hominids, based on fossil and biochemical data.

Blood Groups of Primates, pages 163–177
© 1983 Alan R. Liss, Inc., 150 Fifth Avenue, New York, NY 10011

10

Spontaneously Occurring Agglutinins in Primate Sera

Discovery. 163
Occurrence in Various Species. 165
Classification and Significance 173

DISCOVERY

The discovery, 100 years ago, that sera of animals of one species almost always react with the red cells of other animal species explained why transfusions between animals of two different species usually gave rise to serious hemolytic reactions [84]. The agglutinating and hemolyzing antibodies responsible for these interspecific reactions are presently called heteroagglutinins or heterohemolysins. To determine whether similar reactions occurred within the same species, Landsteiner [85] matched sera and red cells of his laboratory staff, and thus made his historic discovery of the human A-B-O groups. These had two features, quite unexpected at that time: 1) strong agglutination reactions and 2) reciprocal relationship between the isoagglutinins in the serum and the agglutinogens on the red cells. These isoagglutination reactions were thought by Landsteiner to be the source of blood transfusion accidents, which could be avoided by selecting A-B-O–compatible blood donors. Later on, Unger [232] noticed that even when donors were selected on this basis, transfusion reactions could occur, which he ascribed to "minor" isoagglutinins as contrasted to the "major" isoagglutinins, anti-A and anti-B. The first systematic study of isoagglutinins in man, other than those defining the A-B-O blood groups, was carried out by Landsteiner and Levine [91]. They named such agglutinins the "irregular" isoagglutinins because of their infrequent, erratic, and unpredictable occurrence, and the consequent lack of a regular reciprocal relationship with the corresponding agglutinogen on the red cells. They found that the most common irregular agglutinins are related to the subgroups of A, and they also described sera that in retrospect can be recognized as of specificities anti-P and anti-Lewis. Another approach used by Landsteiner and Levine [86] was to carry out the tests at refrigerator temperature, which demonstrated the existence not only of so-called "cold" isoagglutinins, but also of autoagglutinins, which reacted with the individual's own red cells. In fact, it was only after these cold autoagglutinins were removed from the sera that the presence of type-specific cold isoagglutinins could be demonstrated.

A new chapter in the study of isoagglutinins was opened by the discovery of the Rh factor [98]. It was soon realized that the sensitization of an Rh-negative person to the Rh factor could occur either through injection of Rh-positive blood [276] or through pregnancy with an Rh-positive fetus [105]. Thus, it became evident that agglutinins other than anti-Rh could also be produced by isoimmunization. Active search for such agglutinins began and led to the discovery of a large number of previously unknown blood groups, so that at present the term "irregular" agglutinins remains only of historical interest.

Another kind of type-specific antibody belongs to the category of heteroagglutinins and heterohemolysins, which have been described before as determining interspecific differences, but which also may detect individual differences within species other than that of the donor of the serum. The first systematic study of such agglutinins was made by von Dungern and Hirszfeld [33], whose important paper is often overlooked. They investigated the reactions with human red cells of sera from many animals of numerous species, after absorption with human group O red cells to remove nonspecific heteroagglutinins. The reagents thus absorbed commonly gave the strongest reactions with human red cells of group A or B, indicating the presence in the animals' sera of anti-A and/or anti-B agglutinins. In addition to these very prominent agglutinins, there were others giving considerably weaker reactions, which apparently defined numerous other specificities unrelated to the A-B-O groups. The number of serological specificities so defined was so great as to make it possible to recognize the individual from whom every blood sample originated. However, the authors did not follow up this work, probably because they found that the agglutinins, other than anti-A and anti-B, gave weak and erratic reactions that were not readily reproducible. Analysis of their tables indicates that a few of the agglutinins had a preferential activity for human O red cells, suggesting the specificity at present designated as anti-H.

In spite of the interest of spontaneously occurring isoagglutinins and heteroagglutinins for the theory and practice of blood transfusion, and for problems of maternofetal incompatibility, immunogenetics, and the mechanisms of antibody formation, there has been no recent systematic study of this problem, applying the methods developed since the study of von Dungern and Hirszfeld.

OCCURRENCE IN VARIOUS SPECIES

The studies on blood groups in nonhuman primates date back to the original work on rhesus monkeys [97] that led to the discovery of the Rh factor in man [98]. Several publications followed, but the concentrated effort began in the early 1960s with the initial survey of human-type blood groups in numerous species of apes and monkeys. At that time large numbers of primates of various species were tested for the presence of spontaneously occurring antibodies, but usually only the human-type anti-A and anti-B agglutinins were found. Therefore, routine tests for agglutinins were abandoned and efforts were concentrated on the more rewarding study of human-type blood groups and the production of isoimmune sera for simian-type blood groups [140]. In the course of these studies, from time to time, spontaneous isoagglutination reactions besides anti-A and anti-B were observed, but these were weak and were therefore not followed up. In 1973, cross-matching tests were undertaken, for teaching purposes, on randomly chosen baboon blood, and weak agglutination of some, but not all, baboon red cells was noticed. Shortly afterwards, in routine blood group tests on baboons for major surgery entailing blood transfusion, cross-matching revealed a strong unexpected antibody in the serum of a nonimmunized animal. These two incidents led to a more systematic study of agglutinins spontaneously occurring in the "normal" sera of primate animals, and the results were first published in 1976 [210].

Numerous experiments were performed in most of which no agglutination was observed or only occasional weak clumping. For example, in one experiment the sera and red cells of 15 rhesus monkeys were cross-tested. All 225 tests by the saline agglutination method at room temperature showed no evidence of clumping. No reactions were obtained by the antiglobulin method either. In tests against ficinated red cells, however, two sera weakly agglutinated, at room temperature, the red cells of all 15 rhesus monkeys, including those from which they were drawn. At refrigerator temperature, these two monkey sera were found to contain nonspecific cold autoagglutinins of two units titer for ficinated red cells of all 15 rhesus monkeys.

Results of a similar experiment are illustrated by Table 10-1, which shows cross tests on the sera and red cells of ten nonimmunized baboons. These tests were done by the saline agglutination and antiglobulin methods, but not

TABLE 10-1. Results of cross-tests by the antiglobulin method on the sera and red cells of ten baboons

Red cells of baboon:	Sera of baboon:									
	1	2	3	4	5	6	7	8	9	10[1]
1	–	tr	–	–	–	–	–	tr	tr	±
2	tr	–	tr	tr	–	–	–	±	tr	+
3	–	tr	–	tr	–	–	–	tr	tr	+ +
4	–	–	–	–	–	±	–	–	tr	+ +
5	tr	–	–	–	–	–	–	–	tr	–
6	–	–	–	–	–	–	tr	tr	+ ±	tr
7	–	–	–	–	–	tr	–	–	–	+ +
8	–	–	–	–	–	–	–	–	–	–
9	–	–	–	–	–	–	tr	tr	–	tr
10	–	–	–	–	–	–	–	–	tr	–

Tests by the saline agglutination method were all negative.

tr=trace; for definition of all other symbols used in this table and those that follow, refer to footnote, Table 9-1, page 111.

[1]Female baboon.

with ficinated red cells. Since all 100 tests by the saline agglutination method gave negative reactions, only the results of antiglobulin tests are shown in the table. As can be seen, the serum of animal No. 10 contained a type-specific antibody. Titration tests confirmed reproducible reactions of three units titer by the antiglobulin method, with no reactions by the saline agglutination method. In parallel tests, the serum of baboon No. 10 proved to correspond to the simian-type baboon blood group antibody anti-Z^p described previously [139]. It is significant that baboon No. 10 was female.

Table 10-2 shows the titers of the antibody found by a different cross-testing experiment in the "normal" serum of another female baboon, Allison B-358. Noteworthy are the high titers and avidity of some of the reactions. Parallel tests with our standard blood-grouping reagents (compare data in Table 9-27) indicated that the serum of this baboon contained antibodies of two already known baboon simian-type blood specificities, namely, anti-B^p [135] and a second, much weaker antibody, anti-G^p [138]. In conformity with previous experience with the reagent anti-B^p, the titers were highest when measured by the antiglobulin method and considerably lower, though of good avidity, when measured by the saline agglutination method; the serum was nonreactive for ficinated red cells. Strangely enough, subsequent injections of baboon red cells of $B^p + G^p$ specificities failed to stimulate the antibody titers in Allison's serum. On the contrary, the agglutinating activity of the serum continued to decline and gradually disappeared during the 8 months following initial testing.

TABLE 10-2. Results of titration of a type-specific antibody discovered in an apparently normal female baboon (Allison B-358)

Red cells of baboon:		saline agglutination	antiglobulin	ficinated red cell
		Titer of antibody in serum No. 358 by the method of:		
Charlie B	B-343	4	64	0
Lila	B-375	0	0.5	0
Marilyn	B-376	0	0.5	0
Eugene	B-9	0	0.5	0
Jack	B-13	4	32	0
Allison	B-358	0	0	0

Recently, 370 serum specimens of hamadryas baboons were screened for the presence of spontaneously occurring antibodies, and 12 of them were found to contain relatively strong agglutinins, most of them type-specific. It seems of importance that, here again, the specificity of the majority of agglutinins in the normal baboon sera was identical or very similar to the isoimmune standard blood grouping anti-B^p serum. Table 10-3 shows, as an example, reactions of a series of baboon red cells with two normal, antibody-containing baboon sera, compared to those obtained with the standard anti-B^p reagent. It is obvious that the B^p antigen is a particularly strong immunogen among simian-type agglutinogens of the baboon red cells.

A comparable situation appears to exist among the macaques, where the screening of large numbers of serum specimens not only revealed the presence, in some of them, of relatively strong spontaneous agglutinins, but also indicated that the specificities of those antibodies were often very similar or somewhat interrelated. In Table 10-4 are presented the results of cross-matching tests carried out with the sera and red cells of 18 randomly chosen crab-eating macaques.

Only the reactions by saline agglutination are shown because the antiglobulin method failed to improve the results. It will be noted that the sera from all but one of the 18 monkeys agglutinated human red cells. In most cases, the reactions were quite strong for both human A and B cells, indicating the presence of species-specific heteroagglutinins [185]. In some cases, however, the heteroagglutinin in the monkey sera appeared to be of anti-A or anti-B specificity. In addition to the heteroagglutinins for human red cells, about half of the sera contained isoagglutinins reactive for red cells of some but not all of the other crab-eating macaques. Most of the reactive sera contained isoagglutinins reactive preferentially for the red cells of macaques J-168 and J-172, indicating the prevalence of isoagglutinins of a particulate specificity.

TABLE 10-3. Comparative results obtained with the red cells of hamadryas baboons in tests with antibody-containing normal baboon sera and with standard anti-B^P isoimmune reagent

Red cells of baboon:	Reactions (by antiglobulin method) with:		
	normal serum of baboon 12883	normal serum of baboon 15472	isoimmune anti-B^P baboon serum (B-9)
11075	+ +	w	+ +
7341	+ +	+ +	+ +
11699	+ +	+ +	+ +
10544	+ +	+ +	+ +
11690	−	−	−
11692	+ +	+	+ +
15553	+ +	+S	+ +
12866	−	−	−
11072	+ +	+ +	+ +
7351	−	−	−
12883	−	−	−
B-9	−	−	−
11255	+ +	+	+ +
12175	+ +	+ +	+ +
15472	−	−	−

w = weak or doubtful; +S = "one strong."

However, isoagglutinins of other specificities were evidently also present; particularly noteworthy was the considerable polymorphism detected by the serum of monkey J-74. In general, the reactions shown in Table 10-4 were of low titer and avidity, so no attempt was made to use these sera as diagnostic blood-typing reagents.

As the result of cross-matching of the sera and red cells of a batch of bonnet macaques *(Macaca radiata)*, two sera were found to contain spontaneous antibodies with type-specific characteristics [205]. The reactions of both sera with red cells of a series of 38 bonnet macaques are shown in Table 10-5.

The distribution of reactions strongly suggests the antithetical relationship, thus indicating that the antigens detected by the two sera are probably part of the same system. Because of the striking polymorphism demonstrated by these tests, an attempt had been made to improve the titer and avidity of the antibodies by isoimmunization with positively reacting red cells. The efforts, however, proved unsuccessful.

It is noteworthy that many of the "normal" sera of bonnet macaques, although inactive against the red cells of monkeys of the same species, still

TABLE 10-4. Results of cross-matching tests by the saline agglutination method on the sera and red cells of 18 nonimmunized crab-eating macaques (*Macaca fascicularis*)

Blood of:	Sera of crab-eating macaques:																	
	J-1	J-7	J-34	J-36	J-44	J-46	J-52	J-62	J-74	J-86	J-104	J-103	J-152	J-156	J-164	J-166	J-168	J-172
M. fascicularis																		
1 J-1	–	–	–	–	–	–	–	–	+	–	–	–	–	–	–	–	–	–
2 J-7	–	–	–	–	–	–	–	–	+	–	–	–	–	–	–	–	–	–
3 J-34	–	–	–	–	–	–	–	–	+	–	–	–	–	–	–	–	–	–
4 J-36	–	–	–	–	–	–	tr	–	+	–	–	–	–	–	–	–	–	–
5 J-44	–	–	–	–	–	–	–	–	–	–	–	–	–	–	–	–	–	–
6 J-46	–	–	–	–	–	–	–	–	tr	–	–	–	–	–	–	–	–	–
7 J-52	–	–	–	–	–	–	–	–	–	–	–	–	–	–	–	–	–	–
8 J-62	–	–	–	–	–	–	–	–	–	–	–	–	–	–	–	–	–	–
9 J-74	–	–	–	–	–	–	–	tr	–	–	–	–	–	–	–	++	–	–
10 J-86	–	–	–	–	–	–	–	–	–	–	–	–	–	–	–	++	–	–
11 J-104	–	–	–	–	–	+	–	–	+	–	–	–	++	–	–	–	–	–
12 J-130	–	–	–	–	–	tr	–	–	tr	–	–	tr	–	–	–	–	–	–
13 J-162	–	–	–	–	–	–	–	–	tr	–	–	–	tr	–	–	–	–	–
14 J-156	–	–	–	–	–	–	–	–	–	–	–	–	–	–	–	–	–	–
14 J-164	–	–	–	–	–	–	–	–	tr	–	–	–	–	–	–	–	–	–
16 J-166	–	–	–	–	–	–	–	–	–	–	–	–	–	–	–	–	–	–
17 J-168	tr	–	–	–	+	++	++	–	+	++	–	–	++	–	–	–	tr	–
18 J-172	+	–	–	–	+	++	++	–	+	++	–	–	++	–	–	–	tr	–
Man																		
Group A	+	++	+	+	+	++	++	–	++	++	++	++	++	++	–	++	++	++
Group B	–	++	–	–	++	++	++	–	++	++	–	++	++	++	+±	++	++	–

tr = trace.

TABLE 10-5. Comparison of the reactions of two naturally occurring isoagglutinins in the sera of bonnet macaques

Red cells of:	Serum of:		Red cells of:	Serum of:	
	MRA-8	MRA-390		MRA-8	MRA-390
1 MRA 175	−	+	20 MRA 747	−	+
2 MRA 231	−	+	21 MRA 765	w	w
3 MRA 249	−	+	22 MRA 779	−	+
4 MRA 333	−	+	23 MRA 781	+	w
5 MRA 339	+	w	24 MRA 784	+	−
6 MRA 385	−	w	25 MRA 786	−	+
7 MRA 413	+	−	26 MRA 790	−	−
8 MRA 424	−	−	27 MRA 796	+	w
9 MRA 430	+	w	28 MRA 801	+	−
10 MRA 443	−	+	29 MRA 820	w	w
11 MRA 573	−	+	30 MRA 822	−	−
12 MRA 596	+	−	31 MRA 825	−	w
13 MRA 638	−	+	32 MRA 829	−	w
14 MRA 650	+	w	33 MRA 832	+	w
15 MRA 700	w	−	34 MRA 16,091	−	+
16 MRA 720	−	+	35 MRA 16,093	−	−
17 MRA 721	−	+	36 MRA 16,101	−	−
18 MRA 724	−	+	37 MRA 8	−	+
19 MRA 736	+	−	38 MRA 390	+	−

− =Negative; + =positive; w=weak or doubtful.

strongly agglutinated blood of other species of macaques, namely, of rhesus monkeys (Table 10-6). Thus, the sera contained heteroagglutinins for red cells of the phylogenetically closely related species. This is comparable to the regular presence in human sera of heteroagglutinins directed against the red cells of all chimpanzees. It should be noted that the standard isoimmune rhesus sera are suitable for blood grouping of rhesus monkeys and pig-tailed, stump-tailed, crab-eating, and Barbary macaques, but they proved of no value for typing red cells of bonnet macaques. These sera either failed to agglutinate any of the red cells from more than 50 bonnet macaques or agglutinated them all [198].

Table 10-7 gives the results of tests performed with antibody-containing "normal" sera of three pig-tailed macaques *(Macaca nemestrina)* with the red cells of 38 animals of the same species, as well as red cells of a few stump-tailed macaques *(M. arctoides)* and rhesus monkeys *(M. mulatta)*. It was obvious that the specificities of sera Nos. 76140 and 76413 were closely related, showing only minor differences that were comparable to those existing between human anti-A and anti-A$_1$. The third serum, No. 76132,

TABLE 10-6. Results of cross-matching tests by the antiglobulin method on the sera and red cells of 11 nonimmunized bonnet macaques (M. radiata)

Red cells of:	Sera of:										
	8-F	48-F	330-F	385-F	398-F	469-M	471-M	563-M	573-F	650-M	720-M
Bonnet macaques (M. radiata)											
1 8-F	—	—	—	—	—	—	—	—	—	—	—
2 48-F	+±	—	—	—	—	—	—	—	—	—	—
3 330-F	—	—	—	—	—	—	—	—	—	—	—
4 385-F	+±	—	—	—	—	—	—	—	—	—	—
5 398-F	—	—	—	—	—	—	—	—	—	—	—
6 469-M	++	—	—	—	—	—	—	—	—	—	—
7 471-M	+±	—	—	—	—	—	—	—	—	—	—
8 563-M	+±	—	—	—	—	—	—	—	—	—	—
9 573-F	+±	—	—	—	—	—	—	—	—	—	—
10 1650-M	+±	—	—	—	—	—	—	—	—	—	—
11 720-M	—	—	—	—	—	—	—	—	—	—	—
Rhesus monkeys (M. mulatta)											
12	+++	+++	tr	+++	+	—	+++	—	—	++±	—
13	++	+++		++	+±	—	++	+±	±	++	—
14	++	++	tr	+++±	+	—	+++	—	—	++	—

M = Male; F = Female; tr = trace.

TABLE 10-7. Comparison of the reactions of naturally occurring agglutinins in the sera of pig-tailed macaques with enzyme-treated red cells of macaques

Red cells of:	Reactions[a] with the serum of monkey:		
	76140	76413	76132
Pig-tailed macaques *(M. nemestrina)*			
1. No. 70151	+ ±	−	−
2. 71451	+ ±	+ ±	+
3. 74009	+ ±	+ +	−
4. 74257	+ ±	+ +	+ ±
5. 74273	+ ±	+ ±	−
6. 76023	+ ±	+ ±	+ ±
7. 76035	−	−	−
8. 76040	−	−	+ ±
9. 76041	+ ±	+ +	−
10. 76043	−	−	+
11. 76119	+ ±	+ ±	
12. 76126	+ ±	+ +	+ ±
13. 76132	+ ±	+ +	−
14. 76143	−	−	+ ±
15. 76144	+ ±	+ +	−
16. 76157	+ ±	+ +	+ +
Stump-tailed macaques *(M. speciosa)*			
1. MSP266	+ ±	+ ±	+ +
2. MSP356	+	−	+ +
3. MSP363	+	−	+ +
4. MSP367	+	+ ±	+ +
Rhesus monkeys *(M. mulatta)*			
1. Rh-2	+ +	+ +	+ +
2. Rh-6	+	+ ±	+ +

[a]At room temperature.

seemed to have an unrelated specificity. A few comparative tests carried out with red cells of other species of macaques indicated that spontaneous agglutinins present in the sera of the pig-tailed macaques were cross-reactive and able to detect a specificity for which stump-tailed macaques were also polymorphic [204].

Of interest are the tests carried out to detect the presence of autoantibodies in the sera from random nonimmunized apes and monkeys. Our experiments proved that such autoantibodies are prevalent in nonhuman primates and, as in man, are generally nonspecific in that they agglutinate not only the red cells of the animal from which they are derived but also red cells from all other tested animals of the same species. Moreover, as in man, the antibodies react almost exclusively at low temperatures. For the most sensitive results, it is necessary to separate the serum from the clot at body temperature to avoid the absorption of the autoantibody by the animal's own red cells and to perform the actual test at refrigerator temperature. In addition to the low temperature, ficin treatment of the red cells also generally increases the avidity and titer of the reactions.

Table 10-8 shows a sample protocol of titrations of autoantibodies found in the serum of a baboon, B-230, both in a preimmunization sample and in a specimen of blood drawn in the course of isoimmunization. This serum was investigated because it had been found to contain cold autoantibodies. Titrations at body temperature by the most sensitive method using ficinated red cells were uniformly negative. At room temperature only weak reactions were observed, while at refrigerator temperature the agglutination was quite strong. However, even in the cold, the red cells of one of the three baboons chosen at random hardly reacted with B-230's serum. Thus, cold autoantibodies are not all necessarily of a non-type-specific kind.

CLASSIFICATION AND SIGNIFICANCE

The studies of the so-called spontaneous agglutinins in the sera of man and primate animals led to some generalizations concerning classification of those

TABLE 10-8. Titration of autoantibodies in baboon serum against ficinated red cells, showing effect of temperature on the reactions

Tested against red cells of:	Serum dilution before injection					Serum dilution at first test bleeding				
	Undil.	1:2	1:4	1:8	1:16	Undil.	1:2	1:4	1:8	1:16
Serum of B-230 at 37°C										
B-230	−	−				tr	−	−	−	
B-143	−	−				±	±	tr	−	
B-17	−	−				±	±	−	−	
B-336	−	−				−	−	−	−	
At room temperature										
B-230	±	±	±	±	−	nd				
B-143	±	+	±	±	−					
B-17	+	+	±	±	−					
B-336	tr	−	−	−	−					
At 4°C										
B-230	+ +	+ +	+	±	−	+ +	+ +	+ ±	+	±
B-143	+ +	+ +	+	±	−	+ +	+ +	+ +	+	±
B-17	+ +	+ +	+	±	−	+ +	+ +	+ ±	+	±
B-336	±	+	+	+	−	tr	−	−	−	−

nd = not done, tr = trace.

antibodies, their properties, and their role in the mechanism of antibody formation [279]. Natural antibodies occurring in animal sera fall into the following categories.

1. *Cold-reactive agglutinins.* These often, but not always, are nonspecific and act as autoagglutinins. It seems evident that they make up a substantial portion of the naturally occurring agglutinins in human and animal sera. When the red cells agglutinated at 4°C are allowed to warm to room or body temperature, the so-called autoantibodies elute from the red cells, which at the same time separate from one another, and this process can be repeated indefinitely by cooling and warming of the mixture. The fact that the cold autoagglutinins are not specific for the individual's own red cells alone, as a rule, but will similarly clump red cells of all or many animals of the same species, and frequently even of animals of other species as well, raised the question whether the substance in the serum responsible for agglutination is really an antibody. Biochemical studies, however, proved that the substances in question are indeed serum globulins, presumably IgM immunoglobulins.

2. *Agglutinins reactive at room as well as refrigerator temperatures but not at body temperature.* Direct tests as well as those carried out after absorption to render the antisera monospecific have proved that the antibodies in question have particular specificities. The most prominent spontaneously occurring agglutinins are the species-specific antibodies, which are so prevalent that agglutination almost regularly occurs when serum of one animal is mixed with red cells of another animal of a different species. The reactions tend to be most regular and intense when the two animal species in question are far apart taxonomically; on the other hand, agglutination may fail to occur between closely related species—for example, when baboons of different population groups are cross-matched. Another example is the failure in general of normal chimpanzee serum to agglutinate human group O red cells, though the reverse mixture of human serum and chimpanzee red cells almost regularly gives rise to agglutination irrespective of blood group.

The naturally occurring agglutinins that are reactive at room temperature can be fractionated by absorption with selected agglutinable red cells. In this way, or even without absorption, some of these agglutinins can be shown to be type-specific, as well as species-specific, when tested against red cells of more than one species. Of the type-specific agglutinins the most prevalent and prominent are those of specificities anti-A and anti-B and, to a lesser degree, of specificity anti-H. Less regular in occurrence are other type-specific agglutinins, such as anti-A_1, anti-P, anti-M, and anti-Lewis. The agglutinins anti-A and anti-B stand out because of the regularity of their occurrence in man and apes, giving rise to the four classic A-B-O groups.

Even in Old World monkeys, whose red cells are not reactive with anti-A, anti-B, or anti-H reagents, this regular occurrence of anti-A and anti-B holds, the antibody production in that case being canalized by the presence of group-specific substances in the secretions and body fluids instead of on the red cells. What is even more remarkable is that in group A as well as AB humans, the tendency to form anti-A is so great that despite canalization of the immune response in some of these individuals anti-A is actually produced, though only a fraction of that antibody, designated anti-A_1. Similarly, in New World monkeys, which have B-like agglutinogen on their red cells, anti-B is often present but only a fraction of those antibodies that are nonreactive with the animal's own red cells.

Closely related to anti-A and anti-B in regularity of their occurrence are those agglutinins of specificity anti-H. The low-frequency occurrence of anti-H in normal human sera can be ascribed again to canalization of the immune response, because H is almost universally present in human red cells and in secretions. Significantly, potent anti-H is regularly present in the serum of individuals of the so-called Bombay type, who lack H as well as A and B. Moreover, anti-H is a common constituent of animal serum, notably eel serum. The prevalence of anti-A, anti-B, and anti-H among the spontaneously occurring agglutinins of human and animal sera has been ascribed to the prevalence of A-like, B-like, and H-like antigens in microorganisms, plants, and animals, so that exposure to such antigens is virtually unavoidable as the result of inapparent infection or the ingestion of food. The agglutinins in question are therefore presumed to be of heteroimmune origin and appear to be mainly IgM immunoglobulins. Similarly, the other, less regularly occurring type-specific agglutinins such as anti-P, anti-M and anti-Lewis are presumably of heteroimmune origin, and in fact the production of potent anti-P in man has been ascribed to the antigenic action of scolices in echinococcus cysts of infested individuals. It is noteworthy that there is no sharp line differentiating type-specific from species-specific agglutinins and agglutinogens. For example, agglutinogen B is group-specific on the red cells of man, orangutans, and gibbons, but B-like agglutinogens are species-specific characteristics in animals such as rabbit and New World monkeys.

3. *Agglutinins reactive at body temperature.* The final category (and presumably one of the most important) of spontaneously occurring agglutinins are those reactive at body temperature. In man, of course, the agglutinin that is most important from practical and theoretical points of view is anti-Rh, though anti-A and anti-B, as well as a host of antibodies of other specificities, may also be reactive at body temperatures. In general, these antibodies are IgG immunoglobulins and are of isoimmune origin, even though in many

cases the antigen that elicited their production is not always apparent. In the absence of experimental induction of their production by injection of blood and blood transfusion, the usual cause of their formation is maternofetal incompatibility and transplacental isoimmunization by mothers with fetal red cells, not only in man but also in nonhuman primates.

Observations on naturally occurring antibodies support the idea that antibody production is in part adaptive as well as genetically influenced, rather than exclusively one or the other, as is frequently maintained. Biochemically, all the antibodies produced by an individual animal as well as by other animals of the same species are similar, despite differences in their serological specificities. In fact, this observation is being practically applied in the antiglobulin test and for Gm and Inv typing. These biochemical similarities must have a genetic basis, but it does not appear reasonable to claim that the variable portions of the immunoglobulin molecules are similarly exclusively genetically determined. The number of genes required for this would be astronomical, and it therefore appears that it would be more reasonable to postulate an enzymatic adaptive mechanism for antibody production, as proposed by Wiener [244]. As the raw materials for such antibody production, one may postulate the naturally occurring immunoglobulins that react as physiological cold autoantibodies and that make up the first category of the agglutinins described above. Presumably, the constituents of one's own body act as templates used by appropriate enzymes to join together prefabricated sections of the constant portions of immunoglobulins in a form convenient for storage. When later an antigenic stimulus enters the body, this is processed by enzymes that hydrolyze the already preformed and stored cold autoantibodies at appropriate points in the immunoglobulin polypeptide chain, and the resulting fragments are then linked together again by using as "angle irons" individual amino acids or oligopeptides, again by enzyme action under the influence of the processed antigen. Of the antibodies thus formed, anti-A, anti-B, and anti-H stand out because of the ubiquitous nature of the evoking antigens, as already pointed out. In support of this hypothesis is the occurrence of cold autoagglutinins of high titer in lymphomatous diseases, presumably due to reversion of the antibody-producing cells to their primordial condition [252]. Similarly, the appearance of potent cold autoagglutinins was observed in the serum of a chimpanzee being subjected to intensive immunization, presumably a reaction by the antibody-producing cells that serves to provide more raw material for antibody production in response to prolonged antigenic stress. (This may be compared to the production of fetal hemoglobin by patients suffering from severe forms of certain anemias.) The cold autoantibodies appearing under antigenic stress may be considered physiological, as contrasted with the potent cold autoagglutinins found in lymphomatous disease. Whether there are important qualitative differences between the autoantibodies produced under these two different conditions remains to be established.

Blood Groups of Primates, pages 179–221
© 1983 Alan R. Liss, Inc., 150 Fifth Avenue, New York, NY 10011

11

Practical Applications of Blood Group Studies in Nonhuman Primates

Serological Maternofetal Incompatability 179
Homologous Transfusion. 188
Heterologous Transfusion 201
Transplantation 203
Blood Groups as Genetic Markers; Paternity Investigations 206
Seroprimatology 212
Primate Immune Sera as Typing Reagents for Human Red Cells 219

As was the case earlier in man, the discovery and studies of blood groups in nonhuman primates proved to have practical ramifications that went far beyond simple individual differences displayed by antigen-antibody *in vitro* reactions.

SEROLOGICAL MATERNOFETAL INCOMPATABILITY

The pathogenesis of erythoblastosis fetalis in man appears to be well understood thanks to the pathfinding work of Levine and his associates [105], for review see [2]. In the classic case, an Rh-negative pregnant woman becomes sensitized to the Rh factor, usually as a result of transplacental leakage of fetal Rh-positive blood into her circulation, or, on occasion, as the result of parenteral injections of Rh-positive blood. The resulting maternal Rh antibodies readily pass through the placenta into the fetal circulation and then coat the red cells of the fetus, giving rise to manifestations of hemolytic disease. Severe cases are stillbirths of hydropic or macerated fetuses with marked hepatosplenomegaly. Less severely affected offspring are liveborn with *icterus gravis,* which may lead to kernicterus causing severe mental retardation. The mildest manifestation is hemolytic anemia. As is to be expected, the severity of the manifestations is correlated with the degree of sensitization of the mother, i.e. the higher the titer of the maternal antibodies, the greater the stillbirth rate [292]. In exceptional cases, however, clinical disease may fail to occur, even though the mother is strongly sensitized and the baby is Rh-positive [72,246]. In a case described by Wiener and Brancato [246], an infant born of a strongly sensitized mother (antiglobulin titer of 128 units) had almost maximally coated red cells at birth with potent free antibodies in the serum (antiglobulin titer 50 units), yet was hematologically and clinically normal. Over a period of months, the degree of coating of the red cells gradually diminished, as did the titer of free antibodies in the infant's serum, yet at no time did the infant exhibit any jaundice or any other clinical evidence of hemolytic disease. These puzzling cases indicated that the pathogenesis of erythroblastosis is not as simple as it is generally thought to be, and that protective mechanisms probably exist that can interfere with the pathogenetic action of antibodies coating the circulating red cells of the infant. It appeared evident that this important clinical phenomenon deserved thorough investigation in an animal model.

Erythroblastosis fetalis occurs naturally in various animal species [162]. In horses, pigs, and dogs the maternal isoantibodies cannot pass through the placental barrier, but appear in the colostrum and thus affect the incompatible offspring only after birth, when it first begins nursing. A situation more similar to that in man occurs in rabbits; in that species maternal isoantibodies pass through the placenta and produce erythroblastosis *in utero* as demonstrated in a series of experiments by Kellner and Hedal [73–75].

Nonhuman primates appear to be the experimental animals of choice because of the similarity of their fetoplacental situation to that of man [26] and the advanced knowledge of blood groups in several primate animal species.

Spontaneous hemolytic disease *in utero* has been reported in marmosets [53,54], the twin chimerism of which renders the situation more complex. The incidence appears to be higher in offspring of matings between marmoset species, which is comparable to the occurrence of hemolytic disease in mule foals.

Maternofetal incompatability has been observed in a large breeding colony of hamadryas baboons (*Papio hamadryas*) maintained at the Sukhumi Primate Center, USSR [233,236]. Retrospective study of 1,127 pregnancies in Sukhumi baboons showed that incompatible matings yielded "pathologic outcome" in almost 16% of the cases, compared with only slightly over 7% in compatible matings. The frequencies of abortions and stillbirths were also twice as high in pregnancies from incompatible matings as from compatible matings. Mortality up to 1 year of age was significantly higher among offspring of incompatible matings. In 19 cases, spontaneous hemolytic disease of the newborn hamadryas baboons could be directly referred to the sensitization of the mother with the fetal erythrocyte antigens.

Experimental production of erythroblastosis in primate animals was first reported in baboons (*Papio hamadryas*) by Volkova et al. [235,236]. The Russian workers, despite their apparent isolation from the mainstream of modern research, as demonstrated, among other ways, by their incomplete bibliography and the use of roman numerals for the A-B-O blood groups [235], were able to devise an ingenious and successful experimental protocol.

To induce blood group incompatibilities they immunized females with the blood of males, and then mated the females to the males against which they had produced antibodies; the antibody titers were increased by two additional injections during the ensuing pregnancy. Of the twelve pregnancies so produced, one resulted in miscarriage and four in stillbirths, four newborns became severely jaundiced, two had moderate anemia, and one was free of disease. Necropsy revealed severe anemia and one stillbirth, and the remaining three showed erythroid hyperplasia in the bone marrow and nests of erythroblasts in the liver and spleen.

This work was later continued and extended by the same group (for review see [233].) Of 43 blood group-incompatible full-term offspring of experimentally isoimmunized mothers, there were 17 stillbirths and 26 liveborn babies, of which 11 had severe icterus (6 deaths, 2 kernicterus) and 8 had anemia (3 deaths). The remaining 7 had no clinical signs of the disease.

Anemia, hyperbilirubinemia (up to 30 mg%), erythroblastosis of the peripheral blood, erythroid hyperplasia of the bone marrow, extramedullary hematopoieses in the liver, spleen, and lymph nodes, and positive direct antiglobulin test in the affected infants confirmed the identity between this syndrome in baboons and erythroblastosis fetalis in man. It is interesting to note, however, that as many as seven offspring were clinically unaffected. Unfortunately, the authors did not indicate whether the red cells of the unaffected incompatible offspring were coated with maternal antibodies, but they state ". . .we failed to find complete parallelism between the titer of isoimmune antierythrocyte antibodies [in the mother] and the severity of the disease in the fetus."

In addition, among 604 full-term baboons from incompatible but not experimentally isoimmunized parents, there were 15 erythroblastotic infants, 7 of them stillborn; and of the remaining 8, as many as 6 died within the first few days of life, with severe jaundice and/or anemia.

In our own studies, however, we were unable to confirm the above results [201]. Four pregnant olive (*P. cynocephalus*) and hamadryas baboon females (previously isoimmunized and known to produce high-titer antibodies) were hyperimmunized in the first half of pregnancy with the red cells of their breeding mates. All pregnancies resulted in delivery at term of normal infants without clinical symptoms of erythroblastosis fetalis, although in two cases the infants' red cells were moderately coated with maternal antibodies. Similar experiments were carried out with four hyperimmunized female rhesus monkeys (*Macaca mulatta*); again, the hemolytic disease failed to develop even though fetal red cells were maximally antibody-coated. Sullivan et al [227] also reported that rhesus infants whose red cells had been maximally coated with antibodies did not show any clinical signs of hemolytic disease.

In another series of experiments, induction of erythroblastosis was attempted in crab-eating macaques (*M. fascicularis*) [288]. Matings of crab-eating macaques incompatible for their simian-type blood groups were selected. As soon as the existence of pregnancy was confirmed, blood was drawn from each female for preliminary compatibility tests, and the male's red cells, washed three times in saline, were packed and mixed with equal amounts of complete Freund's adjuvant until fully emulsified, and 1 cm^3 of

TABLE 11-1. Results of studies on three pregnancies of crab-eating macaques isoimmunized with the breeding mate's red cells

	Case No. 1	Case No. 2	Case No. 3
Pregnant female:	J-34	J-46	J-86
male partner:	J-1	J-1	J-7
Onset of pregnancy (date)	1/5/71	1/17/71	1/22/71
Injection of father's blood (date)	3/3/71	3/3/71	3/3/71
First test bleeding of mother (date)	4/14/71	4/14/71	4/14/71
Saline agglutination titer	24[a]	8	0
Antiglobulin titer	100[a]	20	200
Bleeding of mother at term (date)	6/6/71	6/23/71	7/2/71
Saline agglutination titer	2	2	0
Antiglobulin titer	40	22	24
Offspring's condition at birth	Normal	Stillborn[b]	Normal
Tests on offspring at birth			
Red cells			
Direct antiglobulin test	Positive	Positive	Negative
Serum (free antibodies)			
Saline agglutination titer	nd	0	0
Antiglobulin titer	nd	8	1
Highest serum bilirubin during first five days of life	4.6 mg/100 cm	nd	nd

[a]Titrations revealed the presence of antibodies of two different specificities, one active in saline and the other in antiglobulin test.
[b]Caused by obstetrical complications.
nd = not done.

the mixture was injected intramuscularly into multiple sites of the female. The pregnant females were bled at the sixth week of gestation and tested for isoantibodies; those with insufficient titers were reinjected intramuscularly with the male partners' red cells emulsified with incomplete Freund's adjuvant. When feasible, test bleedings of the pregnant females were repeated at the third month of pregnancy and shortly before term. Immediately after delivery, the birthweight of the newborn animal was determined and the cord blood collected. Complete blood counts, reticulocyte counts, and icterus index determinations were carried out on the surviving infants at birth and at intervals thereafter. In addition, the newborn infants' red cells were tested for antibody coating by the direct antiglobulin method, using rabbit antirhesus globulin serum.

Table 11-1 summarizes the chronology and results of experiments carried out on three female monkeys who completed pregnancies. As shown here, pregnant females developed potent isoantibodies against the red cells of their mates. The titers of antibodies were relatively high and persisted until term, at which time they were moderately lower. Transplacental passage of antibodies could be ascertained in two of three newborn babies whose red cells were found to be maximally coated with maternal antibodies, as shown by positive direct antiglobulin tests as well as by presence of free antibodies in the infant's serum. Yet, the fetal coated red cells remained intact and no clinical symptoms of intravascular hemolysis were observed.

These and observations reported above indicate that the macaques, as well as some of the baboons that failed to develop clinical symptoms of hemolytic disease despite the presence of maternal antibodies on the offspring's red cell surface, must benefit from some kind of protective mechanism. The nature of this mechanism is unknown, although some explanations have been proposed to account for similar but extremely rare cases described in man. It could be a polymorphic hereditary inhibitor or an occasional physiological state, as, for instance, the adequate hydration of the newborn baby which prevents clumping of coated red cells as shown by *in vitro* tests [292].

In a series of experiments described recently by Stone et al [221] uncoated and antibody-coated red cells of rhesus monkey and cattle were tagged with ^{51}Cr and injected into rhesus monkeys, and their *in vivo* survival was compared by estimating the rate of clearance from the circulation. The results obtained indicated that the antibody coating of the red cells had little or no influence on the shortening of the survival of the injected cells. This in turn signified that "rhesus alloimmune antierythrocyte antibodies do not behave as opsonins and, thus, do not enhance clearance of antibody-coated cells from circulation." It is possible, as Stone et al. point out, that the maternal

rhesus antibodies also do not behave as normal opsonins to facilitate phago-
cytosis, and/or the macrophages simply lack a receptor by which opsonized
cells are normally removed from circulation. This would account for the
absence of clinical symptoms of hemolytic disease of the newborn in monkeys despite the undeniable existence of transplacental immunization and
maximum coating of the fetal erythrocytes.

In contrast to baboons and macaques, two newborn chimpanzees were
described with clinical symptoms of erythroblastosis fetalis and their early
death could be ascribed to transplacental passage of maternal antibodies
directed against antigens on their red cells [286]. A third chimpanzee baby,
born with severe jaundice, a high bilirubin level, and red cells strongly
agglutinated by the mother's serum, was observed more recently. It is
noteworthy, however, that all three cases of transplacental immunization in
chimpanzees occurred in mothers who were previously isoimmunized against
a red cell antigen that, by coincidence, was also present on the red cells of
their breeding mates. Incompatible pregnancies constituted a booster that
caused reapparance of antibodies that remained in the mother's circulation
for some time after delivery. Significantly, in two of those cases the specificity of antibodies in the maternal circulation was related to the R^c antigen,
which is known to be a chimpanzee counterpart of the human Rh_0 antigen
(see page 80). In the third case, the antibody in question was most probably
of anti-V^c specificity. The latter, in turn, was found to be identical to, or
closely related with, human N specificity (see page 69).

At the time of this writing, one of the authors (W.W.S) has been asked to
investigate blood samples of a family of orangutans that included a pregnant
female in approximately the second trimester of pregnancy, her breeding
mate, and their two living offspring. The female's previous pregnancy resulted in a live-born baby that died after 10 days, and the necropsy revealed
changes characteristic of erythroblastosis fetalis, namely jaundice, hemosi-
derine deposited in the liver, spleens, and lungs, and intrahepatic foci of
extramedullary hematopoiesis. Serological tests during that pregnancy were
not performed; postpartum testing consisted of indirect antiglobulin tests in
the mother and her two older babies, all negative.

The tests performed by us included cross-matching of sera and red cells of
all four members of the orangutan family, direct antiglobulin tests as well as
blood grouping tests (with a panel of standard chimpanzee typing reagents).
The initial test on the pregnant female's serum revealed potent antibodies
that agglutinated the red cells of her breeding mate and two offspring but did
not react with her own red cells. The antibodies were mainly of IgG type,
and their titer slowly but steadily increased during the second half of preg-

TABLE 11-2. Titers of antibodies in the serum of a pregnant orangutan female (against red cells of the breeding mate Bongo)

	Method of titration		
Date of serum sample	Saline	Antiglobulin	Enzyme-treated RBCs
December 11, 1980	3	12	128
February 5, 1981	2	16	256
March 2, 1981	6	64	256
April 19, 1981 (postpartum)	6	64	384
April 19, 1981 (postpartum) (against red cells of baby Boris)	0	Over 512[a]	Over 512[a]

[a]No endpoint established. Cells heavily coated.

nancy (Table 11-2). The specificity of the antibodies of the female's serum could not be established, but the results of blood grouping tests, presented in Table 11-3, showed that there was a possibility of maternofetal incompatibility related to an Rh-like or R^c-like antigen. The pregnancy resulted in natural delivery of a full-term live-born male infant with symptoms of severe hemolytic anemia of the newborn: marked jaundice and anemia, thrombocytopenia and normoblastosis, quickly rising bilirubin levels (up to over 400 mg/liter), maximally coated red cells (strongly positive direct antiglobulin test), and free maternal antibodies in baby's serum. Repeated exchange transfusions of maternal red cells resuspended in a neutral medium brought about almost immediate recession of the most severe symptoms, and 1 month after birth the baby was still alive and doing well. Detailed serological analysis of the postpartum samples of the mother's serum, as well as tests on the eluates obtained from the baby's red cells, confirmed earlier suppositions of the relationship of maternal antibodies to the human Rh_0 and chimpanzee R^c blood factors.

Aside from perinatal observations, the importance of maternofetal incompatibility is ascertained from the relatively frequent occurrence of so-called natural antibodies (i.e., antibodies not resulting from deliberate immunization) in the sera of randomly screened primate animals. According to Volkova et al. [236], isoantibodies frequently found in the sera of multiparous macaques resulted from naturally incompatible pregnancies. Our observations seem to confirm that view. Screening the sera of 340 rhesus monkeys maintained for the National Institutes of Health, Division of Research Resources, revealed spontaneous antibodies in about 15% of the samples tested; the great majority of antibody-containing sera were from adult females (unpublished). In a series of 52 bonnet macaques (M. radiata) maintained at the California Primate Research Center at Davis, four contained natural antibodies in their sera; of these, two showing the strongest activity were obtained from multiparous females [205]. Transplacental immunization was most probably the source of strong natural agglutinins found in the sera of several baboons, as well as a possible source of antibodies in the sera of four of 18 nonimmunized crab-eating macaques that we have tested [210]. Bogden and Gray [10] detected widely reactive hemagglutinins in postpartum sera of female rhesus monkeys as late as 45 days after delivery.

We have found spontaneously occurring agglutinins in various Old World monkey species that may act in various ranges of temperatures as isoagglutinins, as heteroagglutinins (both species-specific and type-specific), and even as autoagglutinins. For all practical purposes, the most important are antibodies that are reactive at body temperature, which, as IgG immunoglobulins, readily cross the placental barrier and therefore most often result from maternofetal incompability. Although the strength and avidity of such antibodies are generally only moderate (usually the titers reach 8–16 units by the antiglobulin method), even these relatively weak agglutitinins may cause untoward reactions when transfusion of incompatible blood is given to an animal whose serum contains agglutinating antibodies.

In breeding, the existence of spontaneous agglutinins should be taken into consideration when prospective breeding mates are being selected. Based on the analogy to human erythroblastosis fetalis, one must consider the possibility of the boosting effect of incompatible pregnancy on the titer of preexisting antibodies with devastating results to the fetus. It is therefore recommended that blood grouping, or at least crossmatching, tests be performed between prospective breeders to eliminate incompatible matings, especially when dealing with multiparous females.

Serological screening of pregnant females seems indispensable in cases of intrauterine experiments involving supportive blood transfusions. As shown earlier in this chapter, pregnant females are to be considered high-risk blood recipients because of relatively frequent occurrences of natural agglutinins in their sera.

When agglutinins capable of agglutinating the red cells of the breeding mate are discovered in the serum of a pregnant female, periodic titrations of her serum, particularly in the second half of the pregnancy, are recommended. The rise of the titer of IgG agglutinins may be indicative of maternofetal conflict, and in such cases life-saving exchange transfusion can be envisaged after the infant is delivered with clinical and laboratory symptoms of severe erythroblastosis fetalis. In our own practice, we were able to save an erythroblastotic newborn chimpanzee by exchange transfusion with

TABLE 11-3. Blood groups of the members of an orangutan family with cases of erythroblastosis

Blood and saliva of orangutan:	A-B-O			M-N	Rh-Hr	Simian-type blood groups			Remarks
	Red cells	Serum	Saliva			V-A-B	R-C-E-F	Other	
Bongo (father)	A_1B	None	SecABH	M	Rh-pos	v.O	RCc_1	n^c,O^c,h^c,K^c,T^c	
Sjaan (mother)	A_1	Anti-B	SecAH	M	rh-neg	$V^q.B$	rc_1	n^c,O^c,h^c,K^c,T^c	
Bernardine (offspring I, born 1976)	A_1	Anti-B	SecAH	M	rh-neg	v.O	rc_1	n^c,O^c,h^c,k^c,T^c	
Bosja (offspring II, born 1979)	A_1B	None	SecABH	M	Rh-pos	$V^q.B$	RCc_1	n^c,O^c,h^c,K^c,T^c	
Carin (offspring III, born 1980)	Blood grouping tests not performed								Died at 10th day: gross pathology; *Erythroblastosis fetalis*
Boris (offspring IV, born 1981)	A_1	None	nd	M	Rh-pos	Simian-type blood groups could not be determined; cells maximally coated with maternal antibodies.			Clinical and laboratory symptoms of *Erythroblastosis fetalis*

nd = not done.

thoroughly washed and resuspended mother's red cells. As described earlier in this section, a similar treatment was successfully applied in the case of an orangutan infant born with severe hemolytic anemia caused by transplacental immunization of the mother and sensitization of the red cells of the incompatible offspring.

HOMOLOGOUS TRANSFUSION

The antigenic diversity of the blood of primate animals and the occurrence of hemagglutinating isoantibodies in their sera imply the possibility of transfusion reactions following administration of incompatible blood or early rejection of mismatched transplant, even when both donor and recipient are of the same species. Blood grouping and compatibility testing should become, therefore, standard elements of all procedures that involve administration of blood or transplantation of organs or tissues in apes and monkeys. These include not only biomedical experiments in which primates are used as models for studying transplantation or transfusion related problems, but also all kinds of experiments in which small quantities of blood are administered in connection with studies on infectious, viral, or parasitic diseases, development of vaccines, etc. Here also belong steadily growing numbers of cases in which blood transfusion is applied as a life-saving measure [150] or a method of choice in the treatment of sick animals as well as a supportive procedure in the course of prolonged debilitating experiments or in acute experiments connected with significant loss of blood—as, for example, in large cardiovascular surgical interventions. Serological matching of donors and recipients also proved to be of importance in attempts to use apes and monkeys as organ donors for human patients [161,217] or as partners in heterologous cross-circulation [47,59].

Information concerning isoimmune transfusion reactions in primate animals is scanty. LaSalle and DeLannoy reported [103] that intravenous administration of isologous red cells frequently failed to induce isoimmunization of macaques, or at most yielded only low-titer antibodies. In the study by Lopas and Birndorf [113] only the use of repeated intramuscular injections with adjuvant suspended cells allowed them to produce in rhesus monkeys hightiter agglutininating and hemolyzing isoantibodies capable of causing severe posttransfusion reactions similar to those seen in human patients following administration of incompatible blood. The previously immunized animals were transfused with quantities of ^{51}Cr-labeled rhesus-incompatible blood corresponding to 5% of their blood volume. Major reactions occurred in 5 of 7 animals and consisted of complete destruction of infused cells within 5

TABLE 11-4. Pairing of rhesus monkeys and baboons for cross-transfusion

Species	Pair No.	Identification of animal No.	Sex	Weight at start, kg	Identification of animal No.	Sex	Weight at start, kg	Type and relative number of incompatible specificities
Rhesus	I	Rh-403	(M)	9.0	Rh-1062	(F)	8.5	Simian, 21/35
monkeys	II	Rh-434	(F)	4.6	Rh-898	(F)	4.2	Simian, 11/35
	III	Rh-1072[a]	(F)	5.5	Rh-1078	(F)	5.4	Simian, 5/35
	III-A	Rh-1110[a]	(F)	5.6	Rh-1078	(F)	5.4	Simian, 6/35
	IV (control)	Rh-894	(F)	4.5	Rh-900	(F)	5.0	Simian, 0/35
Baboons	I	B-321	(M)[a]	6.7	B-806	(M)	6.25	Simian 11/45
	I-A	B-842	(F)	6.6	B-806	(M)	6.25	Simian 11/45
	II	B-227	(M)	7.75	B-253	(M)	7.50	Simian 10/45
	III	B-354	(F)	14.4	B-608	(F)	15.7	Simian, 8/45
	IV	B-285	(M)	4.5	B-289	(M)	4.5	Simian 8/45 Human AB/B
	V (control)	B-267	(M)	4.25	B-904	(F)	4.25	Simian 0.45 Human AB/AB

[a]Animal died during the course of experiment from causes unrelated to transfusion of blood, cross transfusions resumed with the new partner three months later.

min, hemoglobinuria, hemoglobinemia, oliguria, anuria (in one case), and disseminated intravascular coagulation syndrome (in three cases). The remaining two animals had mainly extravascular destruction of the transfused red cells and exhibited no coagulation changes.

In our own study [206] five pairs of rhesus monkeys and six pairs of baboons initially received three transfusions of partner's blood at 3-week intervals. The animals were paired by similar weights and by degree of incompatibility of their blood groups, as defined by human A-B-O groups and simian-type red cell specificities. One pair of baboons and one pair of rhesus monkeys, used as negative controls, were identical to each other with respect to all blood groups tested for (Table 11-4). Each animal was given 27–30 ml of blood/kg body weight per transfusion, a volume approximately equal to four units of blood in human transfusion. Test bleedings and titrations of animals' sera were performed at regular intervals. Basic hematologic parameters (red and white cell, platelets, and reticulocyte counts, total and free hemoglobin, hematocrit, and indirect bilirubin) were determined at the same intervals. In the earlier phases of the experiment, survival of transfused red cells was roughly assessed by indirect differential agglutination. The method consisted of the use of potent, specific isoimmune rhesus or baboon typing sera that selectively agglutinated the donor's red cells but not those of the recipient. Agglutination of the posttransfusion samples of blood with such a reagent was considered evidence of the continued presence of donor's red cells in the recipient's circulation; disappearance of agglutination in consecutive samples was considered indicative of total elimination of the transfused red cells.

In the second phase of the experiment, carried out 3 months to 2.5 years after completion of the first series of three transfusions, cross-transfusions were resumed in one pair of rhesus (Table 11-5) and three pairs of baboons (Table 11-6). This time, however, the intervals between transfusions were, by and large, longer, and survival of transfused red cells was estimated by the half-life of ^{51}Cr-labeled red cells instead of the cumbersome an inaccurate differential agglutination method.

All animals except one rhesus and one baboon survived multiple transfusions of incompatible blood without any discernible clinical or hematologic symptoms of acute transfusion reaction, even in the presence in the recipient's serum of relatively potent antibodies against donor red cells. The animals that were lost during the study died shortly after the first or second transfusions from causes unrelated to the administration of blood.

Production of antibodies following initial transfusions was in general slow to start. Only one rhesus monkey responded to the first transfusion with antibodies of significant titer: weak saline agglutinins and medium-strength IgG antibodies appeared in a day 7 sample of blood. The titer of the latter antibodies reached the peak at the time when saline agglutinins disappeared from the serum, i.e. between the 14th and 21st day following the first transfusion, and remained high during the 3 weeks that followed the second transfusion. Of the remaining monkeys, two responded with very weak incomplete agglutinins, first discernible during the third week after the second transfusion of blood, while the sera of all other rhesus monkeys remained inactive throughout the whole course of the three initial transfusions and during the following 5–6 weeks.

Among cross-transfused baboons, only two (B-806 and B-842 in Table 11-6) responded to the first transfusion with antibodies of substantial titer; in others, the appearance of antibodies was delayed until after the third transfusion or did not occur at all during that phase of the experiment.

Among animals that did not produce antibodies even after three consecutive transfusions of large quantities of blood were, as expected, two rhesus monkeys and two baboons presumed to be of identical blood groups, and therefore designated as control pairs. However, when a second series of transfusions was initiated in the control pair of rhesus monkeys after a prolonged period of rest, antibodies appeared in the sera of both animals in a matter of a few days, shortly reaching considerably high titers in one of the monkeys (pair IV, Table 11-5). That finding, together with earlier observations that monkeys exposed to multiple transfusions generally responded slowly and weakly to multiple transfusions, attracted our attention to the importance of the timing of transfusions and particularly to the significance of the length of intervals between transfusions for the promptness and strength of immunological response of the transfused animals. The role of timing in the progress of immunization is demonstrated by the results of serological follow-up of two pairs of rhesus monkeys (including the control pair IV discussed above) and three pairs of baboons who resumed transfusions after periods of rest ranging from 3 months to 2.5 years.

Tables 11-5 and 11-6 show the results of antibody titration in pairs of rhesus monkeys and baboons that completed two full series of transfusions. It can be seen that the animals' responses to the resumed transfusions were stronger, as a rule, than their responses in the course of the initial series of transfusions. In 8 of 10 animals a steep rise of antibody titer was observed during the first and second week following the first boosting transfusion. Further increase of titer took place after the second transfusion of this series. It is significant that, in general, much prompter reactions were noted in animals that had some residual titer of antibodies at the time of resumption of transfusions than in animals whose sera showed no agglutinating activity in pretransfusion samples.

There is no clear evidence whether the relative number of incompatible red cell specificities has any bearing on antibody production in cross-transfused rhesus monkeys. The quickest and strongest antibody response during the first series of transfusions was observed in a monkey that differed from its partner in only 5 of 35 specificities tested for (pair III, Table 11-4). However, its partner, who necessarily differed by the same number of specificities, failed to respond with any antibodies. A similar "asymmetry" of antibody response was observed in pair I, which showed the highest number of incompatible specificities: potent antibodies appeared in the serum of one animal of the pair already during the first series of transfusions and increased further shortly after transfusions were resumed. Antibodies in the partner's serum did not develop until late in the course of transfusions and never reached significant titers. It is noteworthy that both animals that promptly produced high-titer antibodies were of type d^{rh}, while their nonresponding partners were both positive for all graded specificities of the D^{rh} rhesus blood group system (see p. 92). This seems to confirm earlier suppositions that the D^{rh} antigens are more immunogenetic than other rhesus red cell antigens.

TABLE 11-5. Antibody titer and survival of transfused red cells following multiple transfusions of isologous blood in rhesus monkeys

Pair No.	Days after initial transfusion	Amount cross-transfused (ml)	Animal's identification	Titer of serum against donor cells			Survival (days)	Animal's identification	Titer of serum against donor cells			Survival (days)
			Rh-403	Sal	AG	ETC		Rh-1062	Sal	AG	ETC	
I	0	250		0	0	0			0	0	0	
	7	—		0	0	0			1	1	2	
	14	—		0	0	0			0	0	0	
	21	250		0	0	0			0	0	0	
	28	—		0	0	0			0	0	0	
	35	—		0	0	0			1	1	1	
	42	250		0	0	0	Over 21 (DA)		0	4	16	15–20 (DA)
	49	—		0	0	0			0	8	8	
	73	—		0	0	0			0	32	16	
Transfusion resumed 30 months later												
	0	250		0	0	0			0	4	0	
	7	—		0	0	0			3	128	2	
	14	—		0	1	0			2	32	8	
	49	20 (^{51}Cr)		1	1	8	14–16 (T/2)		0	54	0	2–3 (T/2)
	56	—		1	1	8			0	128	10	
	63	—		0	1	1			1	128	16	
	70	—		0	1	1			1	128	8	

Day	IV (control)	Rh-894 Sal	Rh-894 AG	Rh-894 T/2	Rh-900 Sal	Rh-900 AG	Rh-900 ETC	Rh-900 T/2
0	150	0	0		0	0	0	
7	—	0	0		0	0	0	
14	150	0	0		0	0	0	
21	—	0	0		0	0	0	
28	—	0	0		0	0	0	
35	—	0	0		0	0	0	
42	150	0	0	Over 21 (DA)	0	0	1	Over 21 (DA)
49	—	0	0		0	0	1	
73	—	0	0		0	0	6	
Transfusion resumed 3 months later								
0	150	0	0		0	4	16	
7	—	2	4		0	16	32	
14	150	1	3		0	32	64	
21	—	1	3		0	64	64	
28	—	24	12		0	128	256[b]	
42	150	24	12		0	256	256[b]	
43	—	256	256[a]	nd	0	256	256[b]	nd
73	—	6	2		4	256	256[b]	

Sal = saline; AG = Antiglobulin; ETC = enzyme-treated red cell method; DA = differential agglutination; T/2 = half-life of ^{51}Cr-labeled red cells; nd = not done.

[a]Positive direct antiglobulin test.

[b]Warm autoantibodies in the serum, and positive direct antiglobulin test with red cells.

TABLE 11-6. Antibody titer and survival of transfused red cells following multiple transfusions of isologous blood in baboons

Pair No.	Days after initial transfusion	Amount cross-transfused (ml)	Animal's identification	Titer of serum against donor cells			Survival (days)	Animal's identification	Titer of serum against donor cells			Survival (days)
				Sal	AG	ETC			Sal	AG	ETC	
I-A	0	200	B-806	0	0	0		B-842	0	0	0	
	7	—		0	0	0			0	1	1	
	14	200		0	12	24			0	12	4	
	21	—		0	8	6			0	8	3	
	28	—		0	32	32			0	32	4	
	35	—		0	32	0			0	32	0	
	42	200		0	8	12	8–10 (DA)		0	8	4	8–10 (DA)
	49	—		0	8	32			0	16	1	
	72	—		0	32	16			0	32	2	
Transfusions resumed 30 months later												
	0	290		1	2	2			0	3	0	
	7	—		4	96	64			1	128	16	
	14	20 (^{51}Cr)		8	32	48			128	256	256	
	48	20 (^{51}Cr)		4	32	32	0 (T/2)		0	8	0	0 (T/2)
	55	—		8	48	48			0	16	8	
	62	—		8	64	96			0	32	8	
	76	20(^{51}Cr autologous)		6	64	48	14–16[a] (T/2)		0	24	12	14–16 (T/2)
	90	—		6	64	64			0	24	8	
	97	—		6	64	64			0	24	4	
	210	130 + 20 ^{51}Cr		8	24	48	2–3 (T/2)		1	9	1	3 (T/2)
	217	—		24	128	128			1	128	24	
	224	—		24	192	512			1	96	24	
II	0	220	B-227	0	0	0		B-253	0	0	0	
	7	—		0	0	0			0	0	0	
	14	—		0	0	0			0	0	0	
	21	220		0	0	0			0	0	0	
	28	—		0	0	0			0	0	0	
	35	—		0	0	0			0	0	0	
	42	220		0	6	1	15–20 (DA)		0	2	0	Over 21 (DA)
	49	—		0	8	0			0	0	0	
	75	—		0	8	3			0	2	1	

Day	Transfused	B-285 Sal	B-285 AG	B-285 ETC	T/2 (B-285)	B-289 Sal	B-289 AG	B-289 ETC	T/2 (B-289)
Transfusions resumed 30 months later									
0	375	0	6	2		0	1	0	
7	—	64	128	128		8	8	24	
14	—	128	256	256		16	64	64	
48	20 ^{51}Cr	2	96	96	0 (T/2)	4	12	8	0 (T/2)
55	—	4	128	192		8	32	48	
62	—	1	192	128		16	64	64	
83	20 ^{51}Cr autologous	64	256	128	14–16[a] (T/2)	4	24	18	14–16[a] (T/2)
97	—	16	256	128		4	24	8	
104	—	12	256	128		2	24	8	
190	150 + 20 ^{51}Cr	12	128	48	2 (T/2)	1	16	3	3–4 (T/2)
197	—	12	192	192		2	48	24	
204	—	2	512	256		2	64	32	
211	—	16	512	256		4	40	32	
IV									
0	130	0	0	0		0	0	0	
7	—	0	0	0		0	0	0	
14	130	0	0	0		0	0	0	
21	—	0	0	0		0	0	0	
28	—	0	0	0		0	0	0	
35	—	0	0	0		0	0	1	
42	130	0	1	4	15–20 (DA)	0	0	4	15–20 (DA)
49	—	0	1	1		0	0	8	
73	—	0	2	0		0	0	0	
Transfusions resumed 26 months later									
0	220	0	0	0		0	1	0	
7	—	1	32	2[b]		0	16	2	
14	—	0	6	3		0	3	1	
52	20 ^{51}Cr	0	4	1	5–6 (T/2)	0	1	0	14–16 (T/2)
59	—	0	6	2		0	1	0	
66	—	0	6	2		0	1	1	
73	—	0	12	3		0	1	1	

Sal = saline, AG = Antiglobulin, ETC = enzyme-treated red cell method, DA = differential agglutination, T/2 = half-life of ^{51}Cr-labeled red cells, nd = not done. [a]Positive direct antiglobulin test. [b]Warm autoantibodies in the serum, and positive direct antiglobulin test with red cells.

Unlike rhesus monkeys, those pairs of cross-transfused baboons that differed from each other in greater numbers of red cell specificities responded, by and large, earlier and with antibodies of higher titer than pairs of baboons with lesser numbers of incompatible specificities. Here again, however, individual differences in responsiveness were noticed within a single pair of animals, yet no particular red cell specificity could be singled out as responsible for this "asymmetry" of the course of immunization.

The changes in hematologic values observed in the course of two series of transfusions were not significant and were as a rule reversible [58]. The most constant was a transient 10–20% drop in the red cell count, with parallel lowering of hemoglobin and hematocrit values occurring during the first 24 h after each transfusion. It seems plausible that this was a dilution effect caused by infusion of relatively large quantities of anticoagulant added to each unit of administered blood. A drop in the red blood cell count was often, though not always, accompanied by a comparable increase of the total white cell count resulting mainly from the rise of neutrophiles. In most cases, the return to pretransfusion values occurred within 2 weeks after transfusion. Another frequent but not constant finding was some drop in platelet count immediately following transfusion and a temporary increase of reticulocyte count observed most often between the 7th and 14th day after transfusion. Significantly, none of the above changes showed any discernible correlation with the titer of antibodies in animal's serum. Bilirubin levels, tested only during the first phase of the experiment, did not show any abnormal values.

Although there was no evidence of increased intra- or extravascular destruction of incompatible red cells, the survival of transfused red cells was significantly shortened in the presence of antibodies in the recipient's serum. Even allowing for inaccuracy of the differential agglutination as the method for estimating the survival of the transfused red cells, the data in Tables 11-5 and 11-6 suggest that already during the first series of transfusions there appeared a reverse correlation between the lifespan of transfused red cells and the titer of agglutinins in recipient's blood: red cell survival was assessed as normal or nearly normal[1] in recipients with no agglutinins in their sera or with traces of agglutinating activity only. Low-titer antibodies were con-

---

[1]Since, during the first phase of the experiment, intervals between transfusions were of 21 days, survival longer than 3 weeks could not be assessed by differential agglutination. When the donor red cells were found to be present in 20-day posttransfusion samples of recipient's blood, their survival was considered normal or near normal. This was based on estimates by Rowe and Davis [167], who found that erythrocytes of nonhuman primates have a half-life (T/2) of 14 days, as shown by disappearance of ^{51}Cr-labeled autologous red cells. This survival time is about half that found in man.

nected with a minor reduction of the red cell survival rate (15–20 days), while in the recipient animals with substantial titers of antibodies, all transfused red cells seemed to be eliminated within 10 days after transfusion.

The relationship between the antibody titer and the survival of transfused cells was confirmed in the course of the second series of transfusions when the disappearance rate of the chromium-labeled donor red cells was used as the method of the survival estimate. The half-life of ^{51}Cr-labeled donor red cells proved to be shortened to about 4–5 days in the presence of low-titer antibodies, while almost immediate elimination of incompatible donor red cells occurred when agglutinating antibodies of complete and/or incomplete kinds attained significant titers in recipient animal's sera. That the accelerated elimination of donor red cells was mediated by the presence of specific agglutinins could be substantiated by the fact that isotope-labeled autologous red cells reinjected into the recipient's own circulation had normal half-life (see footnote on preceding page).

The curves in Figure 11-1 illustrate the differences in the survival time of the transfused homologous ^{51}Cr-labeled red cells in each of the recipient animals in a pair of cross-transfused rhesus monkeys and a pair of baboons. The half-life of the transfused red cells was assessed when the transfusions were resumed after a prolonged period of rest. The reduced survival correlates well within the increased titer of antibodies (see Tables 11-5 and 11-6).

It is noteworthy that the half-life values of transfused red cells depended somewhat on the volume of the administered blood. As shown in Figure 11-2, elimination of the isotope-labeled red cells from the recipient's circulation was significantly more rapid when the small volume of 20 ml of tagged red cells was injected directly into the recipient animal than when the labeled red cells were first suspended in a larger volume of unlabled donor blood. The half-life of autologous isotope-labeled red cells served here as control.

Observations on incompatible transfusions in rhesus monkeys and baboons have some important practical implications, probably valid for all Old World monkeys. These can be summarized as follows.

1. Transfusions of therapeutic quantities of incompatible isologous blood elicit production of agglutinating antibodies, mainly of univalent type, but do not lead to acute transfusion reactions even when *in vitro* tests indicate a high degree of incompatibility.

2. Production of antibodies is depressed and/or delayed when large quantities of blood are administered at relatively short intervals and the immune system of the recipient is temporarily overwhelmed by the excess of antigen.

3. Preexisting antibodies against donor's red cells (either spontaneous or residual after previous contact with incompatible antigen) intensify the re-

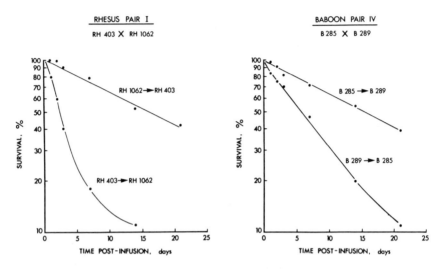

Fig. 11-1. Survival of ^{51}Cr-labeled red cells cross-transfused into recipient animal (rhesus monkey or baboon). Note differences in half-life within a single pair of cross-transfused animals.

sponse to incompatible blood in the form of an earlier and steeper rise of antibody titer.

4. There is no clear evidence that the relative number of antigenic differences between the donor and recipient has a direct bearing on intensity of antibody production. However, there are indications, at least in rhesus monkeys, that some of the red cell specificities may be more immunogenic than others and incompatibilities with respect to these should be given priority when a recipient is being matched with prospective donor(s).

5. Although the mechanism of elimination of incompatible isologous red cells could not be established, there is little doubt that a correlation exists between the level of antibodies in the recipient's serum and the survival rate of the transfused incompatible red cells. High-titer antibodies may be responsible for almost immediate disappearance of transfused red cells, while antibodies of lower titers appear to cause curtailment of survival to one-third of normal half-life.

6. Antibody-related accelerated elimination of incompatible red cells substantiates the need for pretransfusion compatibility testing in all cases when normal survival of transfused red cells is of importance either for therapeutic effect or for success of an experiment.

The effects of incompatible blood transfusion have not been studied in any species of anthropoid apes, though transfusion is occasionally performed in

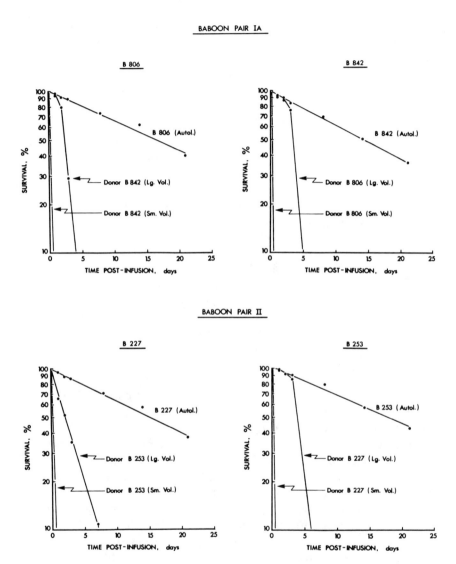

Fig. 11-2. Survival of the ^{51}Cr-labeled red cells cross-transfused into recipient pairs of baboons. The small-volume cross-transfusion was done with 20 ml of labeled red cells of the partner animal. The large-volume cross-transfusion was done with 20 ml of labeled red cells suspended in 130–150 ml of untagged blood. Autologous transfusions of 20 ml of the animal's own blood (^{51}Cr-labeled red cells injected directly into vein) was done as a control.

apes for treatment of severe anemias [150] or as a life-saving procedure for erythroblastotic infants [286].

Routinely, a few prospective donors are preselected from the blood grouping records on the basis of their blood types, and then major and minor cross-matching tests are carried out with the blood of the recipient animal and two or three prospective donor animals.

Since the choice of prospective donors is sometimes limited by immediate availability of animals, it is not always possible to find donors that would be identical with the recipient with respect to all blood groups for which the animals are tested. This is particularly true of chimpanzees, which display a very great variety of blood types. Fortunately, not all chimpanzee red cell antigens are equally immunogenic and, as it is the case with matching donors for human patients, some of the blood group differences can be disregarded.

Because of the presence of the preformed anti-A and anti-B isoagglutinins in chimpanzee sera, A-B-O incompatibility is to be avoided in the first place. Understandably, a group O chimpanzee (with anti-A and anti-B in its serum) should not be given group A blood. Although transfusion of group O blood to a group A chimpanzee is less likely to cause undue *in vivo* reaction, it cannot be excluded that large quantities of anti-A agglutinins contained in group O plasma may bring about some destruction of the recipient's red cells, as has been observed time and again in human tranfusions [62,218,257].

As for other chimpanzee red cell antigens, their role in first-time transfusion is, theoretically, limited to recipients with previous exposure to blood (through experimental immunization or transplacental contact) or with rare spontaneous antibodies of unknown origin. If, however, matching of recipient and donor blood groups is not adequate at the time of initial transfusion, sensitization may occur that would complicate further transfusions if and when they should be necessary. Based on the immunization experiments, we range the chimpanzee red cell specificities in the following order of decreasing immunogenicity: R^c—c^c—V^c—A^c—B^c—C^c—G^c—others.

Preselection of prospective donors takes into account, in the first place, differences within specificities at the top of the above list and tries to avoid donors positive for highly immunogenic specificities when the receeipient is negative for the given blood type.

Cross-matching tests, which precede final selection of the blood donor(s), consist of testing the recipient's serum against red cells of prospective donors (major cross-match). The recipient's own red cells are usually added as negative controls, and also for detection of autoantibodies. The minor cross-match is carried out by testing the donor's serum against the red cells of the recipient. The cross-matching tests are routinely carried out by three agglutination techniques (saline, antiglobulin, and enzyme treated red cell methods) at body temperature and (at least for enzyme-treated red cells) at 4°C. Although weak cold agglutinins (also of autoantibody type) are frequently detected in the sera of nonimmunized primate animals, only high-titer antibodies of this type are considered meaningful. As a rule, the direct antiglobulin (Coombs) test is performed at the same time with the red cells of the recipient animal.

HETEROLOGOUS TRANSFUSION

The use of nonhuman primates in biomedical research occasionally creates situations in which monkeys and apes are to be given human blood, or human patients have to receive the blood of primate animals. In either instance, proper matching of the recipient's and donor's blood is of importance for achieving adequate survival of transfused blood. Under the most frequent conditions, such matching has to take into consideration only two sources of possible incompatibility, namely differences within the A-B-O groups and reactions resulting from heterospecific antibodies. According to Hume et al. [67], the A-B-O incompatibility was the cause of early death of the baboon partner in two cross-circulation experiments with man.

A paper by Goldsmith et al. [59] deals with the possible use of primate animals as partners for cross-circulation with patients in hepatic coma and gives detailed serological follow-up of baboons and rhesus monkeys exchange-transfused with A-B-O compatible human blood. Three group B baboons and three group B rhesus monkeys had their blood gradually replaced by human group B blood. A group AB baboon was, in the same way, exchange-transfused with human group O blood, to which A and B group substances were added in order to neutralize the anti-A and anti-B isoagglutinins. The quality of blood needed to achieve the desired degree of replacement of the animal's blood was calculated from the formula proposed by the authors: $R = (1-s/V)n$, where R = proportion of the animal's own blood remaining; V = blood volume, assumed to be 7% of the total body weight; s = size of syringe; n = number of syringefuls of blood exchanged.

Using that formula, it was estimated that transfusion of a quantity of blood equal to one total volume of animal's blood yielded 64% exchange, while four blood volumes were necessary to achieve nearly complete (98%) replacement of the animal's blood. As established by the method of differential agglutination, the percentage of human red cells in baboon's blood remained stable for 4 days after completion of exchange transfusion, at which time severe hemolysis and jaundice appeared, accompanied by a sharp fall in hematocrit and a steep rise of the white cell count. A quick rise in antihuman antibody titer followed: from negligible values in pretransfusion baboon serum samples to the titer of 256 units and over, 10–12 days later.

In another experiment of the same series, a baboon was cross-transfused with a chimpanzee of the same A-B-O type. The exchange-transfused baboon followed the same course as animals treated with human blood: leukocytosis, hemolysis, jaundice, and anemia beginning at the fourth posttransfusion day. The animal survived the experiment, but developed potent anti-chimpanzee hemagglutinins and cross-reacting antibodies for human O and A red cells. The chimpanzee, on the other hand, showed no clinical ill effects of cross-transfusion but quickly developed high-titer anti-baboon antibodies.

Good tolerance of human blood was observed by Slapak et al. [193] in hepatectomized rhesus monkeys who received up to 1,000 ml of human group B blood during the anhepatic phase, and the transplantation and posttransplantation periods. That massive transfusion of heterologous blood probably resulted in tolerance to human protein.

In contrast to baboon and rhesus monkeys, who had little or no natural agglutinins against human red cells, preformed heteroagglutinins directed against baboon red cells were found in normal human sera, although their titer rarely reached titers of 4–8 units (by the acacia method). In the case discussed by Fortner et al. [47], a patient with acute hepatic failure who was cross-transfused with A-B-O-compatible baboon blood produced antibodies, mainly of IgM type, that not only agglutinated baboon red cells to the titer of 60 units (at peak of immunization) but also reacted avidly, almost to the same titer, with red cells of various other primate species, namely chimpanzee, gibbon, gelada, and rhesus monkey. It is noteworthy that the cross-circulation was without effect on the titer of anti-A isoagglutinins in the patient's serum.

Cross-circulation between man and chimpanzee was attempted by Seigler et al. [191] to save the life of a 3-year-old child in terminal hepatic coma. Since pretransfusion testing showed relatively strong preformed anti-chimpanzee heteroagglutinins in the patient's serum, the group O animal received initial partial exchange transfusion of 2,000 ml of human type O blood. This was followed by 6 h of cross-circulation. The animal tolerated the procedure without any difficulty; it must be assumed, therefore, that the preformed heteroagglutinins in human blood did not cause immediate destruction of chimpanzee red cells. On the sixth day following cross-circulation, the animal underwent a hemolytic reaction manifested by hematuria, hemoglobinuria, and oliguria. The hemolytic crisis was probably the result of the immune response of the chimpanzee to human erythrocytes, but the serological evidence was lacking since the cross-matching tests were limited to samples taken shortly before and after cross-circulation.

Survival of human red cells introduced into the circulation of primate animals was investigated by Castro et al. [18] for the purpose of finding a proper animal model that could be used for *in vivo* study of intrinsically abnormal human red cells, e.g. sickle red cells. Small quantities of ^{51}Cr-labeled human erythrocytes of group O, Rh-positive were transfused to A-B-O-compatible recipient animals (chimpanzees, rhesus monkeys, and baboons), and their recovery was estimated from posttransfusion blood samples. In some instances, nonlabeled human umbilical cord erythrocytes were used instead of isotope-tagged red cells, and their recovery was estimated from post-transfusion blood smears that had been stained by the acid elution method (Nierhaus-Kleihauer-Betke method). The results obtained (summarized in Table 11-7) indicated that the chimpanzees tolerated human erythrocytes better than other primate animals tested: all of the human red cells injected into two chimpanzees circulated 15 min after transfusion and were subsequently removed from the animals' blood with half-lives of 1.6 and 2.4 days. In baboons and rhesus monkeys, over 95% of the human red cells were cleared within 15 min after injection. There was no direct correlation between the titer of antihuman heteroagglutinins in the animals' sera and the survival rate of human red cells.

In a subsequent study [19] chimpanzees were injected simultaneously with ^{59}Fe-labeled human sickle erythrocytes and ^{51}Cr-labeled normal human red cells. Significantly lower recovery and shorter survival of abnormal red cells indicated that sickling of abnormal human erythrocytes takes place in chimpanzee circulation, thus confirming the usefulness of these apes as animal models for sickle cell anemia research.

TRANSPLANTATION

Relatively little is known of the role of blood group compatibility in the survival of allografts in primate animals or for the success of primate-to-man heterotransplantations. The results of renal function studies in baboons with allotransplanted kidneys described by Murphy et al. [151] showed that compatible A-B-O-matched animals survived an average of 26 days, while incompatible A-B-O-matched animals survived an average of 14.6 days. Compatible animals also showed better recovery in postoperative hematocrit depression. Because of the absence of the A and B antigens on baboon red cells, and because of the possibility that their renal tissue was also free of these substances, it was believed that baboon-to-man kidney kidney heterografts would not be adversely affected by blood group mismatches. However, when patients with O blood type received kidneys of group B or AB baboons, they

TABLE 11-7. Serological findings and posttransfusion recovery of human type O RBCs in nonhuman primates

| Recipient | Weight (kg) | Human-type blood group | Preformed anti-human RBC agglutinins (titer)[a] | | | | Human RBCs transfused (ml/kg) | Post transfusion recovery (%) |
			Saline	Antiglobulin	Ficin-treated cells 37°C	Ficin-treated cells 4°C		
Chimpanzee								
Ch-306	9.2	O	—	—	—	6	0.17	100.0
Ch-168	56.0	O	—	—	—	32	0.11	100.0
Rhesus monkey								
Rh-812	6.5	B	—	—	—	4	0.19	5.0
Rh-754	7.1	B	—	12	12	32	0.17	3.7
Rh-804	5.0	B	2	2	2	16	0.26	1.8
Rh-1105	2.9	B	—	—	—	1	0.27	45.3
Baboon								
B-311	6.7	B	—	—	—	2	0.19	1.2
B-319	7.7	AB	—	—	—	4	0.16	0.8
B-808	8.8	B	—	—	1	16	0.16	1.2
B-958	11.3	AB	2	2	2	16	0.13	1.6
B-345[b]	3.5	B	—	0.5	0.5	8	0.38	5.0

[a]The titer is expressed as the reciprocal of the highest dilution of the serum that gives clearly visible microscopically agglutination of the red cells.
[b]Injected with nonlabeled human cord blood RBCs.

fared slightly less well than three other patients who were transplanted with A-B-O-matched baboon kidneys [217].

Are the simian red cell specificities relevant for histocompatibility? The first to raise that question were Balner et al. [4], who noticed that rhesus antierythrocyte reagents often contained antileukocyte antibodies. They could not prove, however, that the antisera actually recognized identical specificities on red as well as on white blood cells; when absorbed with positively reacting red cells, the reactivity of antisera against leukocytes was never significantly reduced. It was assumed, therefore, that the cross-reactivity of hemagglutinating antisera with the white cells resulted from the fact that whole blood was used for immunizing the monkeys. Disparate conclusions were reached by Bogden and Gray [10]. They immunized rhesus monkeys with fresh, viable leukocyte preparations, and after a series of injections they produced antisera that showed strong and reproducible hemagglutinating, leukoagglutinating, and leukotoxic activity in *in vitro* test systems. Results of tests against erythrocyte and leukocyte panels indicated that, in addition to isoantigens peculiar to either red or white cells, there were antigens shared in common by both cell types. Sixty percent of definitely positive leukoagglutinating and leukotoxic reactions were associated with hemagglutinating activity. The authors concluded, therefore, that there was a possibility that rhesus monkeys had a histocompatibility hemagglutinogen system such as the H-2 in the mouse and the R-1 in the rat. Elucidation of such a system in the lower primate would indicate its possible existence in man and provide a most useful tool in solving transplantation compatibility problems. The nature of simultaneous hemagglutinating and antileukocyte activity of the rhesus isoimmune sera was further investigated by Hirose and Balner [64]. A specially designed computer program was used to analyze reactivity patterns of the sera against erythrocytes and against leukocytes and to select proper cells for absorptions. In addition, absorption procedures were refined in that "pure" leukocyte and erythrocyte suspensions were used as absorbing antigens. In that way, one hemagglutinating rhesus reagent was identified as possibly carrying antibodies directed against an antigen shared in common by erythrocytes and the white cells. Despite that, the rhesus red cell antigens appeared unlikely to play an important role in transplantation: skin grafts obtained from donors positive or negative for the erythrocyte specificities against which recipients were preimmunized did not (with one exception) show significantly different survival time.

More recently, we had an opportunity to investigate some of the Balner's rhesus allaoantisera for their hemagglutinating activity. Of the 32 sera tested, eight were found to react strongly and to relatively high titers with monkey red cells, and some of them contained type-specific antibodies, identical with those displayed by certain of our rhesus blood grouping reagents. In a few cases, the alloantisera appeared to have a mixture of antibodies of at least two different specificities, each reactive by a different hemagglutinating technique. The same was true of some of the chimpanzee alloantisera kindly provided by Hans Balner for our analysis. Eight of 21 reagents tested were found to agglutinate selectively some, but not all, of the chimpanzee blood samples tried. Comparative tests carried out side by side with standard chimpanzee blood grouping reagents revealed that hemamgglutinating activity of chimpanzee alloantisera was often related to the specificities of the chimpanzee V-A-B-D or R-C-E-F blood group systems.

As yet, it has not been possible to perform cross-absorptions and titrations to determine whether the ability of the rhesus and chimpanzee alloantisera to agglutinate the red cells of the respective species reflected the presence of identical specificities on the red as well as on the white cells of those animals. A continuation of investigation is certainly desirable to clarify the nature of relationships existing among the red cell specificities and leukocyte antigens of nonhuman primates and the role of the former for histocompatibility.

BLOOD GROUPS AS GENETIC MARKERS; PATERNITY INVESTIGATIONS

Increasing restrictions on the importation of primate animals from their countries of origin stress the importance and urgency of domestic breeding as the sole means of assuring an uninterrupted supply of animals for biomedical research, and in some instances to prevent the disappearance of an endangered species. Since breeding practices commonly used in colonies of primates tend to be built around a few, select animals known for their good reproductive performance, there is a real danger of progressive impoverishment of the genetic patrimony of the captive primate populations To prevent the effects of inbreeding and to preserve the animals' reproductive capacity, an insight into the genetic structure of the population is necessary, and practical means have to be found to monitor the dynamics of the gene pool, both in time and in space. This can be achieved by identifying and surveying some of the traits that are known to be direct products of genes and that are readily accessible to investigation.

Although one could list a number of heritable traits, both physical and biochemical, only some of them meet the requirements of genetic (or chromosomal) markers [189].

By definition, the genetic markers are body components that exhibit heritable variation sufficient to be classified as genetic polymorphisms. Polymorphic traits are only those traits that occur in a population more frequently than would be expected on the basis of recurrent mutations. For example, the so-called inborn errors of metabolism usually do not satisfy this requirement. Among other characteristics that suggest the usefulness of a given trait as chromosomal marker are 1) its discontinuous and unitary nature; 2) detectability in unchanged form, from early stages of life until death of an individual, by relatively simple and easily reproducible techniques, 3) unequivocal mode of inheritance, possibly by individual genes, sharply differentiated from one another, 4) existence of several polymorphic variations at the same locus, and 5) frequencies varying significantly from one population to another.

The genetic markers most commonly used are either molecular or antigenic variants that are either immediate gene products or, at least, final products of relatively short chains of reactions initiated by a gene, so that the chances of extrahereditary factors influencing the appearance of those markers are negligible.

Among numbers of such heritable variants, the components of blood have been the objects of most extensive investigations. From the available literature, which lists over 40 molecular components off chimpanzee blood investigated, only 10 have so far been found polymorphic, thus meeting the basic requirements of genetic markers. Table 11-8 lists those variants, together with two classes of heritable antigenic types of blood cells.

The choice of genetic markers for routine testing should be based on the relative usefulness of a given set of markers and the ready availability of the testing tools (reagents, equipment, adequate controls, and above all experienced personnel).

The usefulness of a system of markers which, in the case of paternity investigations can be mathematically expressed, is by and large measured by the number of polymorphic forms (complexity of the genetic structure) and depends on favorable distribution of those types in the population [280]. As shown in Table 11-8, the white and red blood cell antigen systems of chimpanzees offer the highest degree of polymorphism. Although no data are yet available on the relative usefulness of various genetic markers of chimpanzees, such information recently became available on pig-tailed macaques [22]. In that study, exclusion probabilities were estimated for each of the 12 polymorphic red cell proteins and two independent red cell blood group systems, the A-B-O and D^{rh} graded systems. The latter two, which together offered eight phenotypes, contributed more than half of the cumula-

TABLE 11-8. Blood components as possible genetic markers in chimpanzee

Number of polymorphic types		Authors
I. Plasma		
Immunoglobulins (Gm,Inc)	2	Alepa [1]
Transferrins	8	Goodman [59a]
Group-specific component (Gc)	2	Cleve and Patutschnick [23a]
Alkaline phosphatase	2	Lucotte [116]
II. Red cell isozymes		
Glucose-6-phosphate dehydrogenase (G6PD)	3	Beutler and West [7c] Lucotte [116]
Nicotinamide adenine dinucleotide phosphate diaphorase (NADPH diaphorase)	3	Meera Khan and Balner [120a]
Peptidase A (triallelic)	4	Meera Khan and Balner [120a]
Peptidase C (triallelic)	4	Meera Khan and Balner [120a]
Phosphoglucomutase (diallelic)	3	Meera Khan and Balner [120a]
Phosphogluconate dehydrogenase (diallelic)	3	Meera Khan and Balner [120a]
III. White cell antigens		
Major histocompatibility complex ChLA (A and B loci)	18	Meera Khan and Balner [120a]
IV. Red cell antigens		
V-A-B-D and R-C-E-F blood group systems and sets of unrelated specificities	67	Socha and Moor-Jankowski ([200,202], unpublished data)

From Socha [195].

tive probability of exclusion obtained when all 14 systems were applied (see Table 11-9).

When computing probability of exclusion of paternity, normally it is assumed that the putative father is unrelated to the biologic father. This assumption is likely to be untenable in a situation of a colony of captive animals where the adult males are commonly derived from the same ancestral stock [22]. Therefore, individual probabilities given by Chakraborty (Table 11-9) were computed, first under the assumption that the putative father is unrelated to the biologic father, and then with the assumption that the brother of the biologic father is the male in question. The contention was that these two numeric values set a range for the true probability in most real-life situations.

Practical applications of blood groups for solving problems of paternity in a large colony of rhesus monkeys were presented by Sullivan et al. [226]. Using a battery of 21 blood typing reagents that defined specificities belong-

TABLE 11-9. Exclusion probabilities estimated from 12 polymorphic red cell proteins and two blood group systems in *M. nemestrina* of Malaysia

Loci	Probability of exclusion	
	Suspected male unrelated to father	Suspected male brother of father
Phosphoglucose isomerase	0.1005	0.0503
6-Phosphogluconate dehydrogenase	0.0283	0.0142
Peptidase A	0.0099	0.0049
Peptidase C	0.0485	0.0243
2,3-Diphosphoglycerate mutase	0.0136	0.0068
Acid phosphatase	0.0079	0.0039
Glutamic-oxalacetic transaminase	0.0050	0.0025
Hemoglobin	0.0050	0.0025
Peptidase B	0.0050	0.0025
Phosphoglucomutase 1	0.0266	0.0133
Phosphoglucomutase 2	0.0050	0.0025
NADP-dependent isocitrate dehydrogenase	0.0164	0.0082
ABO type blood group	0.0836	0.0419
D^{rh}-complex	0.1099	0.0549
Cumulative probability of exclusion	0.3832	0.2027

From Chakraborty et al. [27].

ing to 13 independent blood group systems, they tested 1,263 complete families among which 46 cases of disputed parentage were encountered. Of these 35 (76%) could be resolved by assigning the most probable parent on the basis of available blood group records.

Sullivan et al. computed exclusion probabilities for randomly chosen males, for each blood group system separately, and cumulative probabilities when 7–13 blood group systems were tested simultaneously. The former ranged from 1% to 27%, the highest being for the so-called blood group system G, known to have at least four alleles. The cumulative theoretical probability for all 13 rhesus blood group systems was close to 73%. The authors stressed, however, that the actual rate of successful exclusions could be much higher or lower, since the assumptions of Hardy-Weinberg equilibrium and random mating were not valid. Indeed, when exclusion rates were compared in two separate troops, one man-made and the other composed of wild-caught animals, the exclusion rate turned out to be much higher than expected in the former group of animals but very close to the expected probability in the latter. The reason was the high genotypic diversity of breeding males in the man-made troop compared to a rather homogenous group of males among wild-caught monkeys.

Since the power of a given genetic system to exclude paternity depends, among other things, on the number of alleles, it can be expected that in chimpanzees the theoretically achievable probability of exclusion of paternity will be even higher than in the rhesus monkey. Contribution to the cumulative probability should be particularly significant from such highly developed blood group systems as V-A-B-D or R-C-E-F, each having a number of multiple alleles within one locus. Exclusion probabilities for particulate chimpanzee blood group systems and cumulative probabilities of exclusion of paternity of random chimpanzee males have not yet been computed; on several occasions in the past, however, blood groups were successfully applied to solve problems of doubtful paternity and to correct pedigrees of colony-born animals. A few illustrative cases are presented here.

In *Case 1* (Table 11-10), two males were taken into consideration as possible fathers of Tucson and Gretchen, two offspring of the female Martha (Ch-138). As can be seen, the question of paternity could be solved by excluding Oscar as the father of Tucson, and at the same time, by eliminating Walter as the father of Gretchen. Tucson was found to be group RCF, a type that could not have come from its mother or from the male Oscar but could have been inherited from the male Walter. Walter, in turn, had to be excluded as the father of Gretchen, who was a phenotype rc_1, to which neither of Walter's R-C-E-F alleles could have contributed. In addition, Gretchen was typed V.A, while both her mother and the male Walter were of type V.B. Since, on the other hand, specificities rc_1 and V.A were parts of the phenotype of the second male involved, namely Oscar, he had to be assigned as the father of Gretchen.

Case 2 (Table 11-10) constitutes not only another example of exclusion of one of the two putative fathers, but also a rare instance where a male could be positively identified as the father. In this case, the male Tabletop (Ch-467) had to be excluded as the father of the baby Justin, since neither of the male's two R-C-E-F alleles (R^2 or r^2) could have contributed to the baby's genotype. On the other hand, the paternity of the second male, Rufe, was positively demonstrated: both Rufe and baby Justin displayed a very rare variant of specificity R^c (designated R^c_{var} or R^c_v) found in only one male and one female of breeding age among 410 chimpanzees so far tested by us.[2]

Case 3 (shown in Table 11-11) involved four males as putative sires of the baby chimpanzee Jayme, offspring of the female Kissey (Ch-1027). As can

---

[2]R^c_{var} or R^c_v is a low-grade variant R^c (originally designated by Germanic R with superscript c) that regularly gives weak but distinctly positive reactions with all anti-R^c reagents [202].

TABLE 11-10. Two examples of the use of blood groups for solving problems of parentage in chimpanzees

Case 1

	Walter (Ch-168)		Martha (Ch-138)		Oscar (Ch-211)		Tabletop (Ch-467)	
	Phenotype	Genotype	Phenotype	Genotype	Phenotype	Genotype	Phenotype	Genotype
	O	O/O	A_1	A^1/A^2	A^1	A^1/A^2	A_1	A^1/A^1
	MN	MN/Mn	MN	MN/MN	MN	MN/MN	M	Mn/Mn
	$V^{pq}.B$	V^{pq}/V^B	V.B	V/V^B	V.A	V/V^A	v.AB	v^A/v^B
	RCF	R^{CF}/r^{CF}	Rc_1	R^2/r^1	$RCEc_1$	R^{CE}/r^1	Rc_2	R^2/r^2
	H^c	H^c/H^c	H^c	H^c/H^c	H^c	H^c/H^c	H^c	H^c/H^c
	K^c	K^c/K^c	K^c	K^c/K^c	K^c	K^c/K^c	K^c	K^c/K^c
	N^c	N^c/n^c	n^c	n^c/n^c	n^c	n^c/n^c	n^c	n^c/n^c
	o^c	o^c/o^c	O^c	O^c/O^c	O^c	O^c/O^c	o^c	o^c/o^c
	T^c	T^c/T^c	t^c	t^c/t^c	T^c	T^c/T^c	T^c	T^c/t^c

Case 2

	Olga (Ch-22)		Rufe (Ch-114)	
	Phenotype	Genotype	Phenotype	Genotype
	A_1	A^1/A^1	A_1	A^1/O
	M	Mn/Mn	M	Mn/Mn
	v.A	v^A/v^A	v.B	v^B/v^B
	rc_1	r^1/r^1	R_vCF	R_v/r^f
	h^c	h^c/h^c	H^c	H^c/H^c
	K^c	K^c/K^c	K^c	K^c/K^c
	n^c	n^c/n^c	n^c	n^c/n^c
	O^c	O^c/O^c	o^c	o^c/o^c
	T^c	T^c/T^c	T^c	T^c/t^c

Case 1 (continued)

	Gretchen (Ch-248)		Tucson (Ch-285)	
	Phenotype	Genotype	Phenotype	Genotype
	A_1	A^1/A^1	A_2	A^2/O
	MN	MN/MN	MN	MN/MN
	V.A	V/V^A	V.B	V/V^B
	rc_1	r^1/r^1	$RCFc_1$	R^{CF}/r^1
	H^c	H^c/H^c	H^c	H^c/H^c
	K^c	K^c/K^c	K^c	K^c/K^c
	n^c	n^c/n^c	c	n^c/n^c
	O^c	O^c/O^c	O^c	O^c/O^c
	T^c	T^c/t^c	T^c	T^c/t^c

Case 2 (continued)

	Justin (Ch-315)	
	Phenotype	Genotype
	A_1	A^1/A^1
	M	Mn/Mn
	v.AB	v^A/v^B
	R_vCc_1	R_v/r^1
	H^c	H^c/H^c
	K^c	K^c/K^c
	N^c	N^c/n^c
	O^c	O^c/O^c
	T^c	T^c/T^c

For details, see text. Shown are phenotypes of animals involved, as well as their genotypes. The latter were inferred directly from pedigrees of larger families and/or harems, not shown in the table, or were calculated as "most probable genotypes" from the gene frequencies. From Socha [195].

be inferred from the data in Table 11-11, two of the males, namely Bambam I and Geronimo, had to be excluded on the basis of their V-A-B-D groups, since neither of them could have contributed to the type V.O of the baby. The male Kobi, on the other hand, was excluded as the father on the basis of his R-C-E-F type: none of the alleles that made up his possible genotypes could have contributed to the blood type of Jayme. Only the fourth male, Cocoa, was not excluded as the father of Jayme. However, unlike the situation in Case 2, the blood groups of Cocoa as well as those of the mother and of the baby belong to types quite commonly found among chimpanzees, and this precludes definitive identification of Cocoa as the father of the baby Jayme.

SEROPRIMATOLOGY

Another field in which blood groups of nonhuman primates can possibly find practical application is so-called seroprimatology, i.e., study of distributions of blood groups in various geographically or otherwise separated populations, for the purpose of taxonomic identification. In a series of such studies, significant differences were found in distribution of some of the blood types among groups of chimpanzees previously classified as separate subspecies on the basis of their morphological characteristics [130]. The results indicated that blood groups could be a part of the set of parameters used for taxonomic classification of common chimpanzees. In view, however, of the limited numbers of animals tested and the lack of comparative data on feral animals, the available evidence is still insufficient to include blood-grouping tests as a routine element of subdivision of *P. troglodytes*.

Unlike the situation among subspecies of the common chimpanzee, the blood groups of pygmy chimpanzees (*Pan paniscus*) seem to be, in many respects, different from those of *P. troglodytes* and may therefore constitute an independent marker of these rare apes. Blood grouping tests carried out on limited numbers of pygmy chimpanzees showed them all to have blood types that are not or are very rarely observed among common chimpanzees [141,195a]. For instance, all pygmy chimpanzees were found to be group A_1, giving strong positive reactions with anti-A_1 reagents (human absorbed group B serum and *Dolichos biflorus* lectin) that were indistinguishable from reactions obtained with human A_1 red cells. Red cells of group A_1 common chimpanzee (*Pan troglodytes*) usually give reactions of lesser intensity, comparable to human intermediate subgroup $A_{1,2}$. According to recent observations by Yoshida et al. [293a], people of intermediate A type (A_{int}) contain in

TABLE 11-11. Example of the use of blood grouping tests for solving problems of parentage in chimpanzees: Case 3 (for details, see text)

Animal's identification	Blood type	Possible genotype(s)
Putative sire I	A_2	A^2/A^2 or A^2/O
Ch-2019, Bambam I	M	Mn/Mn
	v.AB	v^A/v^B
	rc_1	r^1/r^1 or r^1/r^2
	O^c	O/O or O/o
	n^c	n/n
	H^c	H/H or H/h
	t^c	t/t
Putative sire II	A_1	A^1/A^1 or A^1/A^2 or A^1/O
Ch-2003, Kobi	M	Mn/Mn
	v.D	v^D/v^D or v^D/v
	RCE	R^{CE}/R^{CE} or R^{CE}/R^C
	o^c	o/o
	N^c	N/N or N/n
	H^c	H/H or H/h
	T^c	T/T or T/t
Putative sire III	A	A/A or A/O (not tested for A_1)
Ch-2004, Geronimo	M	Mn/Mn
	v.AD	v^A/v^D
	RCFc	R^{CF}/R or R^{CF}/r or R/r^{CF} (not tested for c_1)
		(not tested for O^c)
	n^c	n/n
	H^c	H/H or H/h
		(not tested for T^c)
Putative sire IV	A_2	A^2/A^2 or A^2/O
Ch-2020, Cocoa	MN	MN/MN or MN/Mn
	V.A	V/v^A
	rc_1	r^1/r^1 or r^1/r^2
	O^c	O/O or O/o
	n^c	N/N
	H^c	H/h
	T^c	T/T or T/t
Female parent	A_1	A^1/A^2 or A^1/O
Ch-1027, Kissey	MN	MN/MN or MN/Mn
	V.A	V/v^A
	Rc_1	R^1/R^1 or R^1/r^1 or R^1/R^2 or R^1/r^2
	O^c	O/O or O/o
	n^c	n/n
	H^c	H/h
	t^c	t/t
Offspring	A_2	A^2/A^2 or A^2/O
Ch-1036, Jayme	MN	MN/MN or MN/Mn
	V.O	V/V or V/v
	Rc_1	R^1/R^1 or R^1/r^2 or R^2/r^1
	O^c	O/O or O/o
	n^c	n/n
	H^c	H/H or H/h
	T^c	T/t

From Socha [195].

TABLE 11-12. Distribution of the V-A-B-D blood groups in common and pygmy chimpanzees

	Reaction with serum of specificity:				Distribution			
					P. troglodytes (n = 503)		P. paniscus (n = 14)	
Designation	Anti-V^c	Anti-A^c	Anti-B^c	Anti-D^c	Number	Frequency	Number	Frequency
v.A	−	+	−	−	94	0.1869	0	0.0
v.B	−	−	+	−	47	0.0934	0	0.0
v.D	−	−	−	+	8	0.0159	14	1.0
v.AB	−	+	+	−	103	0.2048	0	0.0
v.AD	−	+	−	+	27	0.0537	0	0.0
v.BD	−	−	+	+	21	0.0417	0	0.0
V.O	+	−	−	−	32	0.0636	0	0.0
V.A	+	+	−	−	85	0.1690	0	0.0
V.B	+	−	+	−	50	0.0994	0	0.0
V.D	+	−	−	+	14	0.0278	0	0.0
Rare types:								
V^q.A	$+^a$	+	−	−	3	0.0060	0	0.0
V^q.B	$+^a$	−	+	−	2	0.0040	0	0.0
V^{pq}.A	$+^b$	+	−	−	3	0.0060	0	0.0
V^{pq}.B	$+^b$	−	+	−	9	0.0179	0	0.0
V^{pq}.D	$+^b$	−	−	+	4	0.0079	0	0.0
V^{pq}.O	$+^b$	−	−	−	1	0.0020	0	0.0

[a]Positive reaction with two out of three anti-V^c reagents.
[b]Positive reaction with one out of three anti-V^c reagents.

their plasmas a unique blood group, GalNAc transferase, that is different from those found in blood of A_1 or A_2 people. It would be of great interest to explore this phenomenon in common and pygmy chimpanzees and see whether group A of these two species of chimpanzee can be differentiated on an enzyme basis also.

While *P. troglodytes* can be either M or MN, all pygmy chimpanzees proved to be type M. In addition, the M antigen of *P. paniscus* differs qualitatively not only from the M of human erythrocytes but also from the M-like antigen detectable on the red cells of the common chimpanzee. Striking differences among common and pygmy chimpanzees have also been observed within blood groups of the V-A-B-D system, which is, as explained earlier in this book, the chimpanzee extension of the human M-N blood group system. Table 11-12 gives comparative distribution of the V-A-B-D blood groups thus far observed in *P. troglodytes* and in *P. paniscus*. As can be seen, of 16 V-A-B-D types already identified on the red cells of common chimpanzees by means of chimpanzee immune sera, only one, namely v.D, was found in all pygmy chimpanzees. The type v.D could be considered, therefore, a species-specific trait of *P. paniscus*, and one could venture a hypothesis that the occurrence of the rare v.D type common chimpanzee is the result of more or less recent hybridization with pygmy chimpanzee.

Another striking difference between blood groups of the two species of chimpanzee concerns the distribution of types of the R-C-E-F system, shown in Table 11-13. While in direct tests with panels of chimpanzee immune sera at least 24 various R-C-E-F types were recognized on the red cells of common chimpanzees, none of those types appears in pygmy chimpanzees. All 14 pygmy chimpanzees so far blood-grouped were classified as $R_{ab}CE$ type, an irregular variant not observed in *P. troglodytes*. Detection of the $R_{ab}CE$ antigen on the red cell of a chimpanzee could therefore be regarded as conclusive for classifying this animal as pygmy chimpanzee.

Comparative data on the distribution of various blood groups among various species of Old World monkeys, particularly among various species of macaques, indicated that blood types of a monkey can be used as one of several parameters facilitating taxonomic classification of the animal [124,125,127]. In Table 11-14, for example, are presented frequencies of positive reactions obtained with rhesus isoimmune sera of various specificities in tests with red cells of six species of macaques. As can be seen, there are significant differences among frequencies with which some of the specificities appear in various species of monkeys. Allowing for the relatively small number of animals of each species tested so far, the fact that some of the specificities do not occur in certain species of macaques, while other specificities were found to be present in all animals of the given species, suggests that such blood types may have some value when morphological parameters alone are insufficient to definitely classify the animal to one or another species. Such blood groups may also be of importance for estimating the degree of hybridization in populations of monkeys inhabiting overlapping geographical ranges.

Striking differences in distribution of red cell specificities among various species of macaques can be put to use for determining genetic characteristics of offspring resulting from experimental egg transfers. In such an experiment, the mature egg is surgically removed from the uterus of a female monkey and implanted in the uterus of a female of another species, whose estrous cycle is properly synchronized with that of the donor female. The recipient is then impregnated with the male's sperm.

TABLE 11-13. Distribution of the R-C-E-F blood types in common and pygmy chimpanzees

Designation	Reaction with serum of specificity:						P. troglodytes (n = 475)		P. paniscus (n = 14)	
	Anti-R^c	Anti-C^c	Anti-E^c	Anti-F^c	Anti-c^c	Anti-c_i^c	No.	Frequency	No.	Frequency
rc_1	−	−	−	−	+	+	107	0.2253	0	0.00
rc_2	−	−	−	−	+	−	1	0.0021	0	0.00
(rc)	−	−	−	−	+	nd	16	0.0337	0	0.00
rCc_1	−	+	−	−	+	+	1	0.0021	0	0.00
rCF	−	+	−	+	−	−	5	0.0105	0	0.00
$rCFc_1$	−	+	−	+	+	+	35	0.0737	0	0.00
$rCFc_2$	−	+	−	+	+	−	2	0.0042	0	0.00
($rCFc$)	−	+	−	+	+	nd	2	0.0042	0	0.00
Rc_1	+	−	−	−	+	+	81	0.1705	0	0.00
Rc_2	+	−	−	−	+	−	26	0.0547	0	0.00
(Rc)	+	−	−	−	+	nd	34	0.0716	0	0.00
RC	+	+	−	−	−	−	2	0.0042	0	0.00
RCc_1	+	+	−	−	+	+	4	0.0084	0	0.00
RCc_2	+	+	−	−	+	−	2	0.0042	0	0.00
RCF	+	+	−	+	−	−	21	0.0442	0	0.00
$RCFc_1$	+	+	−	+	+	+	35	0.0737	0	0.00
$RCFc_2$	+	+	−	+	+	−	26	0.0547	0	0.00
($RCFc$)	+	+	−	+	+	nd	18	0.0379	0	0.00

RCE	+	+	+	−	−	−	8	0.0168	0	0.00
$RCEc_1$	+	+	+	−	+	+	15	0.0316	0	0.00
$RCEc_2$	+	+	+	−	+	−	4	0.0084	0	0.00
(RCEc)	+	+	+	−	+	nd	4	0.0084	0	0.00
RCEF	+	+	+	+	−	−	10	0.0210	0	0.00
$RCEFc_1$	+	+	+	+	+	+	1	0.0021	0	0.00
$RCEFc_2$	+	+	+	+	+	−	1	0.0021	0	0.00
(RCEFc)	+	+	+	+	+	nd	3	0.0063	0	0.00
Irregular forms:										
$R_{var}Cc_1$	+[a]	−	−	+	+	+	2	0.0042	0	0.00
$R_{var}CF$	+[a]	−	−	+	−	−	6	0.0126	0	0.00
$R_{var}CFc_1$	+[a]	−	−	+	+	+	2	0.0042	0	0.00
$R_{ab}CE$	+[b]	+	+	−	−	−	0	0.00	14	1.00
$R_{var}CEF$	+[a]	+	+	−	−	−	1	0.0021	0	0.00

nd = not done.

[a]Weak reactions with all three anti-R^c reagents.

[b]Positive reaction with only one of three anti-R^c reagents.

TABLE 11-14. Distributions of red cell specificities (defined by rhesus isoimmune sera) in six species of macaques

| Antiserum | | Percentage of positive reactions with the red cells of: | | | | | |
Identification	Specificity	Barbary macaques n=32	Stump-tailed macaques n=30	Bonnet macaques n=52	Rhesus monkeys n=340	Stump-tailed macaques n=89	Crab-eating macaques n=62
1. Rh-1	Anti-A^{rh}	nd	100.0	100.0	94.71	71.9	91.8
2. Rh-2	Anti-B^{rh}	100.0	0.0	100.0	90.29	4.5	70.2
3. Rh-216	Anti-B^{rh}	100.0	100.0	100.0	90.29	96.6	53.3
4. Rh-4	Anti-C^{rh}	100.0	100.0	100.0	100.0	98.9	100.0
5. Rh-6	Anti-D^{rh}	100.0	0.0	0.0	67.06	86.5	89.6
6. Rh-942	Anti-D$_3^{rh}$	100.0	33.3	nd	43.53	80.9	82.6
7. Rh-212	Anti-D$_2^{rh}$	0.0	0.0	0.0	34.41	39.3	0.0
8. Rh-924	Anti-D$_2^{rch}$	87.5	30.0	nd	43.75	39.3	0.0
9. Rh-218	Anti D$_1^{rh}$	0.0	0.0	nd	33.60	34.8	5.5
10. Rh-224	Anti-D$_1^{rh}$ + Anti-X	0.0	0.0	0.0	34.41	34.8	0.0
11. Rh-228	Anti-D$_1^{rh}$ + Anti-X	0.0	0.0	0.0	53.53	79.8	0.0
12. Rh-930	Anti-D$_1^{rh}$	0.0	0.0	nd	61.76	94.4	85.7
13. Rh-3	Anti-F^{rh}	100.0	100.0	0.0	39.41	96.6	83.3
14. Rh-178	Anti-G^{rh}	0.0	100.0	100.0	84.71	76.4	100.0
15. Rh-926	Anti-J$_3^{rh}$	0.0	0.0	nd	29.12	68.5	100.0
16. Rh-946	Anti-L^{rh}	0.0	0.0	nd	29.12	100.0	0.0
17. Rh-936	Anti-M^{rh}	100.0	0.0	nd	47.65	68.5	28.6
18. Rh-321	Anti-N^{rh}	100.0	100.0	nd	96.76	100.0	100.0
19. Rh-932	Anti-N^{rh}	100.0	100.0	nd	96.76	100.0	21.4
20. Rh-796	Anti-J$_1^{rh}$	100.0	0.0	nd	29.41	14.6	100.0
21. Rh-938	Anti-J^{rh}	0.0	0.0	nd	43.23	9.0	100.0

nd = not done.

The success or failure of the fertilization of the transplanted egg can be assessed by genetically characterizing the offspring that may result from such a pregnancy. We were asked recently to provide such genetic proof by blood-grouping tests in an experiment involving a crab-eating macaque egg donor, a rhesus female egg recipient, a rhesus sperm donor, and the newborn baby. The tests were performed with a set of rhesus and baboon isoimmune antisera that were known to give contrasting results with the blood of rhesus monkeys and cynomolgus monkeys (crab-eating macaques), namely high frequency of positive reactions with one species of red cells but low frequency with blood of another species. Some of the reagents can be recognized among the rhesus antisera listed in Table 11-14.

Table 11-15 gives the most significant results of tests carried out with red cells of the four monkeys: the egg donor, the egg recipient, the male parent, and the offspring. As can be seen, the newborn baby's blood showed the same patterns of reactivity as the blood of the rhesus male parent and that of the "foster" mother. This was particularly evident with the antisera that gave positive reactions with the red cells of the egg donor monkey, but uniformly negative reactions with the blood of the three remaining animals involved in the experiment (the sera anti-G^{rh}, anti-J_3^{rh}, anti-J^{rh}), or vice versa: negative reactions only with the cynomolgus egg donor blood, but positive with red cells of the three remaining animals (the sera anti-B^{rh}, anti-D_3^{rh}, anti-F^{rh}, anti-M^{rh}, anti-N^{rh}, anti-B^p). In general, when the sera reacted differently with red cells of rhesus monkeys and differently with blood of crab-eating macaques included in the tests, the baby's blood followed the patterns of the rhesus blood reactivity rather than that of the crab-eating macaques. The question of maternity was definitely answered by the results of tests with antisera of the D^{rh} series, namely two anti-D_2 and an anti-D_1 reagent: all three reagents agglutinated red cells of the baby as well as those of its "foster" mother, but failed to react with the red cells of the egg donor and with those of the male parent. This indicated that the D^{rh} specificities of the baby's red cells must have been inherited from the rhesus female egg recipient. In conclusion, the results of blood grouping tests not only suggested that the baby inherited the red cell specificities of the type encountered in rhesus monkeys, but they also implicitly excluded the crab-eating macaque egg donor as the genetic mother.

PRIMATE IMMUNE SERA AS TYPING REAGENTS FOR HUMAN RED CELLS

Increasing demand for reliable high-titer anti-A and anti-B typing reagents, use of which already exceeds thousands of liters annually, creates the need

TABLE 11-15. Blood grouping tests in an egg transfer experiment

Red cells of:	Reaction with antiserum:													
	Anti-B^{rh} Rh-216	Anti-D$_3^{rh}$ Rh-942	Anti-D$_2^{rh}$ Rh-212	Anti-D$_2^{rh}$ Rh-924	Anti-D$_1^{rh}$ Rh-218	Anti-F^{rh} Rh-3	Anti-G^{rh} Rh-178	Anti-J$_3^{rh}$ Rh-926	Anti-J$_1^{rh}$ Rh-796	Anti-J^{rh} Rh-938	Anti-M^{rh} Rh-936	Anti-N^{rh} Rh-932	Anti-B^p B-9	Anti-C^p+Q^p B-11
Cynomolgus female (egg donor)	–	–	–	–	–	–	pos	pos	pos	pos	–	–	–	w
Rhesus male (parent)	pos	pos	–	–	–	pos	–	–	–	–	pos	pos	pos	pos
Rhesus female (recipient)	pos	pos	pos	pos	pos	pos	–	–	pos	–	pos	pos	pos	pos
Offspring	pos	pos	pos	pos	pos	pos	–	–	pos	–	pos	pos	pos	pos
Controls														
Rhesus monkey I	pos	–	–	–	–	–	–	–	–	–	pos	pos	pos	pos
Rhesus monkey II	pos	pos	pos	pos	pos	pos	–	pos	pos	pos	–	pos	pos	pos
Cynomolgus monkey I	–	–	–	–	–	pos	pos	pos	pos	pos	–	–	–	w
Cynomolgus monkey II	–	pos	–	–	–	–	pos	pos	pos	pos	–	–	–	w
Baboon I	nd	nd	nd	nd	nd	nd	–	nd	nd	–	pos	nd	pos	pos
Baboon II	nd	nd	nd	nd	nd	nd	–	nd	nd	–	pos	nd	pos	pos

– = negative; pos = positive; w = weak; nd = not done.

for new sources of supply of those reagents. Several primate species seem to be perfectly equipped to become donors of commercial quantities of the A-B-O reagents, thus supplementing the current main source of anti-A and anti-B typing sera, i.e. hyperimmune sera obtained from human volunteers. Group A chimpanzee hyperimmunized with purified group B substance will yield antibodies of anti-B specificity of titers comparable to those of human hyperimmune sera. Depending on the size of animal, up to 500 ml of plasma can be harvested weekly by means of plasmapheresis, so that large quantities of good reagent, practically free of contaminating species-specific antibodies, could be obtained from one animal in a matter of months. In a similar way, anti-A and anti-B antisera can be produced by hyperimmunization of baboons of group B and A, respectively. Heterospecific antibodies in such sera can be eliminated by diluting them out without significant weakening of type-specific agglutinins.

As described earlier in this book, in the course of immunization of chimpanzees with chimpanzee or human red cells, antisera could be produced that not only would distinguish individual differences among chimpanzee red cells, but also, in some instances, would give type-specific reactions with human blood. Two such sera were developed recently, one with a human-reactive component of anti-Rh$_0$ (D) specificity (see page 79) and the other of anti-Mi^a (Miltenberger) specificity (see page 67). Both sera contained relatively weak species-specific heteroagglutinins that were easily eliminated either by dilution or by absorption with human red cells of the proper type. The type-specific antibodies left behind after absorption were of specificities identical to the homologous reagents of human origin; they also had titers comparable to those of commercial antisera of the respective specificities.

It seems possible that a systematic screening of all primate typing sera currently available will disclose many more examples of reagents with type-specific reactivity against human red cells. Some of them, particularly those of rare or very rare specificities, may find practical application as reagents for routine use in blood-grouping tests on human red cells. Such may be the case of the chimpanzee anti-Miltenberger serum, which, as shown by comparative tests with anti-Mi^a reagents of human origin, can satisfactorily replace human antisera whose supply depends on the erratic availability of the rare human donors.

Blood Groups of Primates, pages 223–237
© 1983 Alan R. Liss, Inc., 150 Fifth Avenue, New York, NY 10011

12

Prospects Offered by the Study of Blood Groups of Nonhuman Primates

Comparative Study of Blood Groups in Populations and Evolutionary
  Mechanisms 223
Types of Blood Factors in Living Species 224
Paleosequences and Neosequences 227
Genetic Polymorphism and Speciation 228
Causes of the Maintenance of Polymorphism and Its Variations in Populations
  of Nonhuman Primates 230
Role of Sequential Redundancies in the Diversifying Evolution of Species 236

COMPARATIVE STUDY OF BLOOD GROUPS IN POPULATIONS AND EVOLUTIONARY MECHANISMS

The path opened by A.S. Wiener and his collaborators showed itself to be useful for illuminating certain aspects of evolutionary mechanisms by providing a fuller understanding of the genetic model of the human blood group systems. The close relationships between simian blood groups and human blood groups is not surprising if one considers the biological kinship of man and monkeys. For a long time this kinship has struck anatomists and physiologists, for whom the characteristics of the related species varied more quantitatively than qualitatively (the brain of the baboon and that of man are constructed on the same pattern; what differs is the relative development of the elements that compose it). This overall similarity is also found at the molecular level when the structures of the same proteins (for example, enzymes) are compared in the different species of primates.

Most of the information that codes the peptide chains follows a certain basic outline; the passage from one species to another is demonstrated only by some small modifications that throw light on the nature and the direction of zoological kinships. And what can be observed on the human level is only the end of a long evolution, the stages of which are revealed by the structures retained in the previous groups. By the same token, it should not be considered that the blood groups as we know them today in man appeared "ex nihilo." Their elaboration must have been as long and as laborious as that of organs as complex as the eye of vertebrates or the cerebral neocortex of the higher primates. These organs, which existed well before us, were prepared by a certain number of transformations occurring on earlier evolutionary levels. They did not achieve the human type at once. Like them, the immunological systems had to undergo a long phylogenic preparation; they came into being by stages. From a basic genetic model, present in the common root, they evolved in a direction special for each lineage. Thus, their study in living primates can help to explain their genesis and their structure, as the study of the brain of the prehominids helps to explain the genesis and structure of the human brain. *This is why the comparative immunohematology of primates is an irreplaceable approach to the understanding of the blood group systems in man;* just as paleontology throws light on the origin and formation of the cranium and of the brain of our contemporaries, which would otherwise have remained amazing and mysterious organs.

TYPES OF BLOOD FACTORS IN LIVING SPECIES

For a long time, although it was not yet known that blood group factors belonged to a definite genetic system, they were classified according to their distribution in the taxonomic classes. Thus, the following categories were distinguished:

1. The *heteroantigens* present in many species, some phylogenically remote from each other. They define the *heterotypes*. Present simultaneously in numerous taxonomic groups, heteroantigens were thought to correspond to the general information characteristic of higher taxonomic levels.

2. The *paraantigens*, which define *paratypes* [173,180]. They are presumed to be characteristic of several species occupying adjoining taxonomic classes. As such, they are attributes of narrower taxonomic units. In this category, for example, are certain Rhesus factors: c (hr'), which is found in man, where it is polymorphic, and in some anthropoid apes, where it appears to be monomorphic; the D (Rh_o) factor of human red cells and its equivalent, R^c, characteristic of the red cells of anthropoid apes, etc.

3. The *isoantigens*, which correspond to isotypes characteristic of a single species (or of very closely related species—for example, the subtypes of the D^{rh} system in rhesus monkeys and in pig-tailed macaques) and are therefore of more recent origin.

All these blood groups have been identified by immunological methods; they therefore could be detected only in those cases in which mutations resulted in the appearance of individuals who lacked the antigen and thus were capable of making specific antibodies. The series of alleles that occupy the same locus gave rise to the fourth category, *alloantigens*.

Blood factors that remained monomorphic escaped the investigators. In this way, immunology reveals a certain number of polyallelic genetic systems that "tag" a part of the hereditary patrimony. They are very useful, as we shall see, for labeling linkage groups as a prerequisite for a genetic inventory of the human race. Immunological methods, however, are useless for detecting the monomorphic loci, i.e. those that did not mutate [176]. These methods cannot inform us about the extent of the polymorphism (frequency of the polymorphic loci in relation to the total number of loci) either in the individual (*coefficient of heterozygosity: H*) or in the population (*coefficient of polymorphism: P*). We shall see as we go on how these difficulties were overcome by the methods of enzymology.

If immunohematology is ineffective in revealing the activity of the loci that did not experience mutation, it nevertheless permits an evaluation of their general role by comparing representatives of two or more species, instead of two individuals of the same species, who are identical as to the monomorphic trait. The method amounts to searching not for the effect of *intraspecific polymorphism* (which does not exist in such loci) but of *interspecific polymorphism*. For this purpose one resorts to heteroimmunization. This consists of immunizing a member of one species (recipient) with the blood of a member of another species (donor). Often, the recipient reacts by producing antibodies active against antigens present in all the members of the donor species. By this procedure, the "immune conflict" is shifted from the taxonomic level of species to a higher taxonomic level (genus, family, order, etc.).

The extent of the reaction depends on the number of different proteins in the donor and the recipient, i.e. on the *phylogenic distance* that separates the two species. When the donor and the recipient are taxonomically far apart, the recipient reacts against numerous antigens, some of which are present not only in the donor species but also in the members of closely related species (para-antigens). The antibodies produced in that way will not only have a species specificity (against the donor species) but also a larger specificity (against numerous species related to the donor species). And this specificity will be all the larger as donor and recipient are taxonomically more distant.

In contrast, if donor and recipient are very close, the number of different proteins is lower and the antibody produced can be strictly specific for the donor species. This is what K. Landsteiner called the "*immunological per-*

spective," which was demonstrated in practice by Uhlenhut's test, used in legal medicine to determine the human or animal origin of blood stains. This test consists of producing antihuman antibodies by injecting human serum into an animal. If a rabbit is immunized, the antihuman antibody obtained has quite a wide specificity: it precipitates not only proteins of human origin but also those of the anthropoid apes and even tailed monkeys of both the Old and New World types (i.e., all of the suborder Anthropoidea). When an antihuman antibody is produced by immunizing a baboon or a macaque, the reaction of the antibody is really strong only with human proteins. This reaction is even more specific if the gibbon (anthropoid ape) is the recipient of human protein [112]. As for the chimpanzee, which in many respects is the primate closest to man, it seemed incapable of producing the antihuman antibody; possibly there is not a big enough difference between the structural genes of the two species for a pronounced immunization to be triggered in one in relation to the other. This was also our experience [112]. (Nevertheless, A.C. Mourant reported that he had produced an antihuman globulin by the immunization of a chimpanzee with pooled human sera (personal communications).

The immunological perspective varies with the phylogenic distance that separates the donor and the recipient. This can be represented by an acute angle as shown in the following diagram:

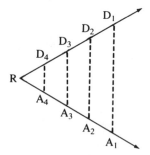

The recipient is represented by a point R situated at the origin; the corresponding donors are at points D_1, D_2, D_3 and D_4; the further these are from R, the greater is the phylogenetic difference between the donor and the recipients (R-D represents their taxonomic distance). The base DA of the triangle corresponds to the zone covered by the specificities of the antibodies produced. The "immunological perspective" can be seen to shrink as D becomes closer to R (it is much broader for D_1 than for D_2, for D_2 than for D_3, for D_3 than for D_4). For two very similar species (e.g., man/chimpanzee) it can be almost nil.

PALEOSEQUENCES AND NEOSEQUENCES

The different immunogenetic systems of primates known today do not in fact demarcate the evolutionary plateaux of the particular narrow taxonomic units, and it seems unjustified to regard them as simian and human systems. The majority of them have a far older origin, having already existed in the common phylum that produced the lineages now alive and coexisting with man. *The phylogenic transformations are to be sought within each system,* where fractions can often be identified that mark certain stages of evolution. The immunogenetic systems have retained traces of evolutionary stages as have the brain and the eye.

The distinctions between heteroantigens, paraantigens, isoantigens, and alloantigens should be rethought in terms of genetic systems rather than separate blood factors.

Thus, the study of the relatively complex blood systems (such as M-N and Rh) frequently makes it possible to distinguish several types of sequences.

1. First, very old sequences are encountered that were present in our common ancestor and that have passed without any great change into all the following lineages. These are *paleosequences* [177–179]. For example, this is the case for a common fraction of the M and N factors, of the D and R^c factors, and of the c (hr') factor, which are found in man and in several anthropoid apes. All these antigens are proof of an ancient evolutionary plateau that preceded the splitting of the initial stock into daughter species [186,187].

2. Other factors or parts of factors are found in only a single species or in a small number of closely related species. These are *neosequences* [177]. They emerged when the secondary lineages split off from the primordial stock. Each one evolved in a parallel fashion but in its own particular way. Examples for the M-N system are the S/s factors in man and the $A^c/B^c/D^c$ factors in the African anthropoids; for the Rhesus system, the E/e factors in man and the $C^c/E^c/F^c$ series in the same anthropoid apes. A peculiarity of these factors is that they all have a corresponding factor in related lineages. All belong to the same genetic set, a fraction of which has evolved independently from the rest.

The appearance of neosequences is probably brought about, on the molecular level, by two possible mechanisms: a duplication (or multiplication) of the same chromosome segment (see redundancies, p 234), each newly formed segment subsequently following a different evolutionary path in various lineages; or a new splicing of the DNA taking place between the *exons* (fractions of DNA translated on the mature RNA) and *introns* (fractions of DNA which, instead of being transcribed into RNA, are excised and expulsed in the course of transcription). The introns, therefore, constitute an abundant reserve of the genetic material, which, passing from the silent stage (introns) into an "expressive" stage (exons), may play an important role in evolution [182]. This mechanism, which seems to explain quite well the evolution of blood groups of primates, was also recently invoked to construct the genetic model of the Rh system of man [57] and to trace the evolution of the globin family of the vertebrates [109].

3. Other factors have been discovered that seem particular to certain species and do not appear to be connected with blood groups present in related species. It is possible that they depend on information sequences that appeared at the moment of speciation; but it is also possible that their connection to already known "human" systems will one day be revealed [177].

GENETIC POLYMORPHISM AND SPECIATION

Analysis of the blood systems found in several species often demonstrates polymorphisms fairly comparable to each other. For the loci corresponding to the paleosequences, common to several groups, the same alleles are encountered (but sometimes one of the pair may be lost). Such is the case, for example, of the A-B-O system, which in its secretory expression has scarcely varied from the New World monkeys up to man. This is true for all the species currently living, which often display polymorphic systems that "span" several related lineages [184].

These data demonstrate that the *birth of a new species depends more on the global and slow transformation of an entire population than on the sudden change occurring in a single individual or a single couple.* Some of the followers of Darwin, now known as "mutationists" (in particular deVries [1901] and later Goldschmidt [1940]), attributed speciation to the appearance of an individual, the carrier of a singular anomaly, which he transmitted to his descendants and then, due to crossings-over, to the entire group. The existence of this "hopeful monster," which would carry all the expectations of the lineage, has never been demonstrated. It is hardly compatible with the transmission of the entire polymorphism of the founder species to the daughter species. In fact, a single individual can only give that which he carries in his genotype, i.e. a tiny part of the genetic variety of the group from which it comes. The charting of individual speciation has recently been taken up again by some cytogeneticists, who see in a chromosome modification occurring in a single subject the origin of a new species [37,61]. This is the *adamic theory of speciation.* Without being able to exclude completely the possibility of such a phenomenon, it seems more plausible to look for the cause of speciation in the transformation of a whole group, progressively isolated from the parent species but still retaining high polymorphism. The birth of a species involves a series of new adaptations and can have an effect only on a population that has a genetic inheritance sufficiently vast and varied to reply to new alternatives, i.e. to colonize a new niche different from that in which its ancestors lived [175,182].

In general, things happen in the following way: a population is living at the periphery of the distribution area of the species, where ecological conditions are at the limits of survival, and where it is exposed to maximum selective pressure. At the same time, due to its marginal and distant position, it has minimal opportunity for gene exchanges with the neighboring populations. If this group has a genetic inheritance sufficiently rich to establish new combinations that are better adapted to these marginal environmental conditions, and if it is sufficiently isolated so that sexual exchanges with the neighboring populations do not challenge the newly acquired factors, a new niche can be colonized. Each new niche being more distant from the parent population, the genetic separation from the parent species becomes so great that it results in the sterility of mutual crosses.

Only at this moment can the chromosome modifications intervene. The role of these alterations in the sexual *isolation* of a group is probably important, but an anomaly of the karyotype only offers a selective advantage if it appears in a population already widely separated from the mother population. The majority of these chromosome modifications occurring in a homogeneous group that is well balanced with its environment are endowed with an unfavorable characteristic because they cross the sieve of meiosis with difficulty. They are quite rapidly eliminated by selection. This sorting explains the usually rigorous constancy of the karyotypes inside the same species, despite the fairly high frequency of chromosome accidents that appear in each generation.

In contrast, when two populations are separating (i.e., when one of them attempts to colonize a new niche), anything that fosters their reciprocal moving-away may be a favorable event, because it diminishes the competition between the two separating groups and consequently increases the resources of each. Thus, from a certain "critical speciation threshhold," the chromosome anomaly is no longer eliminated but is retained by selection if it is compatible with the survival of the individual [182].

In this way, *chromosome modifications do not provoke speciation: they confirm it and render it irreversible.* The remarkable work of Bernard Dutrillaux [37] and the studies carried out in the field on animal groups in the process of splintering into new species (the *Papio*, for example) demonstrate the likelihood of a mechanism of *population speciation*, found in all zoological or botanical groups in which this phenomenon has been studied (see, in particular, Ayala [3] for *Drosophila* of Meso-America).

CAUSES OF THE MAINTENANCE OF POLYMORPHISM AND ITS VARIATIONS IN POPULATIONS OF NONHUMAN PRIMATES

Not all species present the same level of polymorphism for a given genetic system. The A-B-O system, for example, has preserved the three principal alleles in some species, whereas others have two, and in still other species only one allele has been preserved.

While the macaques *M. mulatta*, *M. fascicularis*, *M. nemestrina*, and *M. radiata* carry all three A-B-O mutations (although in the last species the O phenotype has not actually been encountered, the existence of gene O could nevertheless be inferred by gene frequency analysis), it seems that still another species of macaques, the stump-tailed macaque (*M. arctoides*), has only group A, thus resembling marmosets (*Colithrix*) of the New World. However, it is possible that the missing A-B-O mutations do exist in that species but were not accounted for because of their low frequency and the small number of animals studied. For example, it was believed for a long time that *all* rhesus monkeys (*Macaca mulatta*) belonged to group B, which was therefore considered a species characteristic of these monkeys. It was only recently, when blood samples were obtained from monkeys originating from new, previously unexplored geographical areas, that individuals having group O, A, and AB were found (compare Table 5-1) and that hereditary transmission of those "new" mutations could be confirmed by family study [197].

One may ask what causes the observed differences in the distribution of genes among the species? And why have they not all maintained the three mutations? Several causes may be involved.

1. *Genetic drift often has a notable effect on numerically small groups.* It can explain the difference in the frequency of the blood groups observed in a single species for populations geographically far apart and perhaps undergoing differentiation toward autonomous races. This situation is well known in the West African baboons [5]. Nevertheless, to arrive at the total loss of certain alleles and the establishment of a single one (which thus becomes a "public factor") an extremely strong process of drift is required. Such a situation has a chance of occurring only in populations whose numbers are very low, a fact that can considerably reinforce the consequences of contin-

gent phenomena. On the other hand, in large groups, the loss of a few individuals carrying a specific allele does not have a marked effect on the whole species, as the flow of interpopulational gene exchanges sooner or later reestablishes the genetic polymorphism within each population. The changes due to drift can be felt to a maximum degree only if a single individual (pregnant female) or a single couple isolates itself and is on the point of leaving the population. This is the *founder effect*.

It is obvious that the single founding couple could leave to the following generations only a tiny fraction of the gene pool of the population from which it comes.

The loss of certain genes observed in a few species (or, at least in some populations, because no species of nonhuman primate has yet been the subject of a complete genetic inventory) could be connected to severe genetic drift linked to the small size of the founding group [182].

2. *The second cause of polymorphism is linked to the variety of the environment,* i.e. the variety of the selective pressure. Naturally, this factor has an effect in so far as the genes considered are not selectively neutral. If the genes were indeed neutral or nearly so, their distributions in various populations would be purely accidental. We shall return to this important question of neutralism later in this chapter.

Let us retain the hypothesis of genes endowed with a definite adaptive value. Genetic inheritance permits the species to adapt to a certain ecological niche. The niche is, in short, only the exterior projection of the possibilities of the genotype. A very close relationship exists between the diversity (polymorphism) of the hereditary patrimony of a population and a variety (extent) of the environment, both spatial and temporal, that it is able to exploit (and that constitutes its ecological niche). Monkeys are more specialized than *Homo sapiens*. Because of its intelligence, which very early gave it the possibility of offering not only organic but also cultural responses to the pressures of its environment, the human species has avoided embarking on a one-way path of specializing evolution. It has remained undifferentiated and genetically highly polymorphic [178].

In contrast, nonhuman primates are far more dependent on a given environment that they rarely leave. Their ecological niche, variable from one species to another, is always far narrower than the human niche. Although their organic specialization is lower than that of the majority of other mammals, they could have lost a part of their genetic polymorphism in the course of specializing evolution. In fact, studies by enzymologists have demonstrated that the genetic polymorphism of nonhuman primates (and even of those species connected to a relatively narrow ecology) is as widespread as the genetic polymorphism of the human species.

We have already mentioned the H and P coefficients that are commonly used to evaluate this polymorphism. H (level of heterozygosity) represents the percentage of polymorphic loci in relation to the total number of loci in a given animal; and P (level of polymorphism) the percentage of polymorphic loci in relation to the total number of loci in a given species. These two coefficients are easily calculated from the results provided by electrophoresis. Indeed, this technique permits the determination of a certain number of proteins (of the blood of vertebrates, or the hemolymph of invertebrates) that are all genetically controlled. Some proteins are identical in all the animals studied, i.e. they are encountered always in the same migratory position in electrophoresis and they correspond to a monomorphic locus. Others can show different migration speeds, which vary from one subject to another, corresponding to as many mutations; they are genetically polymorphic [115,117]. Table 12-1 shows the different values of H and P in some primate species [114,116].

TABLE 12-1. Measurements of the polymorphisms in some species of nonhuman primates

Species	Rate of polymorphism (P)	Rate of heterozygosity (H)	Rate of allelism (A)
Chimpanzee			
(Pan troglodytes)	0.181	0.022	1.24
Macaques			
Macaca fuscata	0.10	0.01	
M. cyclopis	0.24	0.04	
M. mulatta	0.37	0.09	
M. fascicularis	0.41	0.10	
Baboons			
Papio papio	0.40	0.031	1.56
P. anubis		0.018	
P. cynocephalus		0.028	
P. hamadryas		0.055	
P. anubis	0.17	0.04	1.05
P. hamadryas	0.22	0.05	1.05
Vervet monkeys			
Cercopithecus ethiops ethiops	0.230	0.030	
C. e. sabeus	0.230	0.021	
C. e. pygerythrus	0.192	0.029	

After Lucotte [114, 116].

While the value of *P* is more or less comparable in all species (which demonstrates that all have more or less the same variety and therefore the same genetic richness), that of *H* varies considerably from one group to another. These variations seem to be connected to the social structure. In species that live in relatively strict harems, *H* is low. Inside each such population based on family structure, some endogamy occurs; here the variation appears to be more interpopulational than intrapopulational, for a given period. In other populations, the intrapopulational variations seem to predominate. However, in all cases, the interpopulation flows (either in time or in space) are probably sufficient to maintain a high level of genetic variety of the group, as demonstrated by the high values of *P*.

3. *This demonstrates that genetic polymorphism is somewhat linked to the social structure of the population.* Groups that consist of numerous individuals with frequent sexual exchanges with neighboring populations will be more polymorphic than groups that consist of a small number of subjects living closed in on themselves, in a state of endogamy. (It is known that interpopulation flows are mainly responsible for maintaining the gene polymorphism of each population.) Moreover, certain structures (a harem with a single male having virtual exclusivity of fecundation) are less polymorphic than others (groups with many males and multiple exchanges). Moreover, life in small isolated groups totally modifies the conditions of selection related to diseases.

In fact, for a pathogenic organism to maintain itself in a population, it must possess sufficient demographic concentration, below which the probability of contaminations among individuals is too low for the endemic disease to persist. In addition closed, small populations quite often escape large epidemic waves. Therefore, it is in the large, open populations, such as the human race, that diseases will persist and result in epidemics. It is currently accepted that a relationship exists between different blood types and sensitivity to certain infectious agents [147]. Thus, it is possible that pathological aggressions constitute an important selective element, at least in relation to the A-B-H factors.

It was mentioned earlier that the A, B, and H antigens, or very similar antigens, were widespread in nature and in particular in the microorganisms that are pathogenic for mammals.

"Antigenic competition," which was first described during pregnancy (fetomaternal conflict), is not limited to the reproductive phase; it can be found throughout life, in situations that expose the individual and the population to the A-, B-, and H-like factors of external origin. An organism will defend itself much better against an attack if, in its immunological equipment, it lacks structural antigens that are identical or similar to those carried by the infectious agent. An environment rich in all kinds of endemic diseases may exercise very varied forms of selective pressure capable of maintaining an extensive polymorphism in a population. Depending on the nature of the germ involved, the O individuals or the A and/or B individuals are preferentially eliminated. As even the severest epidemic never affects everyone among the susceptible subjects, such a population will still preserve some blood group polymorphism.

What would happen if these varied pressures did not exist? It is probable that, in a very tightly closed population escaping all selective pressures of an infectious kind and therefore immunologically left to itself, O individuals would have a tendency to eliminate A and B subjects through the generations due to a process of transplacental alloimmunization (on condition, however, that the initial frequency of the *O* gene is higher than the frequency of *A* or *B*, which is the case in all known human populations). Such a process might explain the preponderance of type O in human populations living in an isolated state (high mountains, deserts, peripheral zones of continents) as well as the surprising monomorphism of the Amerindians of South America [149].

It is known that America was populated by groups from the Far East between 10,000 and 50,000 years ago and that, if we can judge from the populations that still live in these regions, they must have been quite rich in A and B factors. However, all the Amerinds, apart from a few tribes of the West of the United States, belong to group O. Arriving in America, a new continent until then unpopulated by man and therefore free of the epidemic diseases of the Old World, these individuals lived in isolated little bands and had little or no chance to spread, in this new environment, the "ancestral pathology" that they had carried to the New World. In the absence of epidemics that would have caused a diversifying selective pressure, fetal-maternal selection acted in favor of group O, which resulted in gradual elimination of the A and B groups.

The extreme sensitivity of the Amerinds to the pathological agents imported from Europe and Africa, which caused so much damage after the conquest of the continent and is still observed in pure-bred groups, constitutes an argument in favor of this theory [183].

In this case, the concurrence of conditions is such that various endemic diseases can be permanently maintained (numerous viral reservoirs, likelihood of contamination, etc.). On this endemic background, occasional epidemic episodes are superimposed. However, they never affect all the individuals. In this way, even the variety of the selective pressures of pathological origin that, in this case, sweeps down on the population becomes by itself a factor contributing to the maintenance of a significant genetic polymorphism.

The infectious "landscape," bacterial, viral, and parasitic, is an integral part of the ecological niche. Here again, the variability of the niche must be accompanied by parallel genotypic changes in the population that occupies it.

4. *The possibility of neutralism, another mechanism that can contribute to maintaining polymorphism, is diametrically opposed to the previously described mechanisms.* It is based on the assumption that certain genes are devoid of all selective value and remain selectively neutral. Their transmission is therefore random. This theory, supported by the Japanese geneticist Kimura [78–82], has lately been taken up by several authors. Without denying the existence of selection, the adherents of neutralism assume, in the first place, that it plays a negative role by eliminating only the truly harmful mutations (Mullerian selection). But the mutations, by and large, remain neutral, and their distribution is random. Kimura and his school advance a number of arguments to support this view.

a. The mutations are much more frequent in the molecules with a secondary role in cellular physiology than in those with essential functions. In other words, the most important molecules show no tendency to evolve.

b. When one takes into consideration the relatively complex molecules, such as hemoglobin for example, one sees that the mutations affect mostly the zones that have little or no functional value. On the contrary, the active sites always remain intact.

c. The same is true, according to Kimura, on the populational level. The animal groups that live in an almost unchanging environment (caves, ocean depths) are as polymorphic as those that live in a constantly changing environment (high- and low-tide areas, for example).

In fact, Kimura assumes that polymorphism tends to increase with neutrality. Functional constraints, on the contrary, favor monomorphism. It is when those constraints diminish that the rate of mutations that are retained reaches its highest level. This can explain the apparent paradox shown by the higher primates. This group, of recent origin, has evolved very quickly toward the hominid branch that led to man. The process of hominization took between 5 million and 7 million years, which is to say a very short time. However, this truly "revolutionary" movement, on the phenotypic level, is characterized by relative molecular immobility. As we mentioned before, very few mutations differentiated man from chimpanzee. In fact, from the eukaryote stage on, the changes that took place involved new combinations and rearrangements rather than new mutations.

ROLE OF SEQUENTIAL REDUNDANCIES IN THE DIVERSIFYING EVOLUTION OF SPECIES

The comparative study of blood groups in populations reveals the *important role played by the multiplication of the same chromosome segment (redundancy) in the evolutionary mechanism.* This process, conceived in 1935 by Bridges [14], was clarified by Ohno [154] and further developed by Ruffié [178].

When the most recent developments in immunohematology are considered, we are struck by the importance and frequency of the multilocus models that are encountered both in the erythrocyte group systems (M-N, Rh, Kell) and in those of histocompatibility (HLA). This convergence is not accidental; it must result from the very mechanics of the evolutionary process. Comparison of the blood group systems in man and in nonhuman primates allows us to envisage the following scheme.

The paleosequences, defined earlier (D and R^c, hr' for the Rhesus system, N^V, N, M, He, Mi(a) for the M-N system, salivary A and B for the A-B-H system, etc.), must have been formed at a certain evolutionary plateau and have persisted without any change in all the resulting lineages (perhaps because they are indispensable for the biological equilibrium of the cell).

These paleosequences underwent a certain number of redundancies (multiplications of the same chromosomal segment), leading to the neosequences that in the Rhesus system, for example, correspond to the E/e antigens of man and $C^c/E^c/F^c$ of the African anthropoids. In the M-N system, the neosequences are the S/s factors in man and the $V^c/A^c/B^c/D^c$ factors in chimpanzees, etc. In the present state of our knowledge, analysis of these systems allows us to trace back the common stock of man and African anthropoid apes that must have lived in the middle of the Miocene era, more than 25 million years ago.

Can we hope to go back even further into the past and establish phylogenic relationships with earlier lineages? Perhaps. For example, in the Rhesus system it is likely that an ancestral factor exists, perhaps an antigenic precursor of the patterns common to the human Rh_0 and the R^c of the anthropoid apes. This factor can be associated with the common ancestor of the Old World monkeys and the Hominoidea living in the middle of the Eocene epoch, 60 million years ago. Later, the locus responsible for its synthesis would have triplicated in the root lineage of anthropoid apes and man. Contrary to the beliefs of Race and Sanger, this triplication was not simultaneous with hominization but must have happened long before, since several Rh-related loci exist in the chimpanzees, gorillas, and gibbons. The product of a single primitive locus, whose existence was postulated by Wiener but never unequivocally demonstrated, still remains to be discovered in the preceding taxonomic groups. The factor in question probably does not correspond to the LW factor but constitutes the last missing link in the phylogenic history of the Rhesus blood group system.

This process of chromosome redundancies seems frequent in the course of evolution; it is encountered in the composition of the hemoglobins, haptoglobins, immunoglobulins, and a good many other active molecules. If it is so widespread, it is because it has been retained by natural selection and because it offers a certain advantage. Today, it seems evident that these redundancies offer new material that can mutate and pass through the sieve of selection without challenging the functioning of the ancestral gene, which itself remains unchanged. Thus, they permit all trials, all evolutionary attempts, without modification of the basic genetic information that controls the fundamental cell processes. It is "adventure without risk," since the evolutionary process can function from these neosequences without touching the paleosequences from which they result, and which are perhaps indispensable to the life of the cell (hence their stability throughout the multiple evolutionary stages). From this available material diversifying evolution can proceed. It enables a primitive stock to splinter into specialized groups, each one capable of occupying a new ecological niche without bringing the fundamental structure of the lineages into jeopardy. They ensure multiple variations on a common theme and could facilitate the "explosion" of certain groups into adaptive dispersions. Moreover, this mechanism can explain, to a certain degree, increasing enrichment and complexity of the genetic equipment of the species, at least from the eukaryote stage.

The exact forms of evolution are difficult to follow. However, the comparative study of blood groups in primates offers a unique tool because it retraces, throughout the millennia, the establishment of the principal blood group systems today encountered in man. In addition, this approach, which was for a long time ignored and neglected, can help us achieve a better understanding of the structures and relationships of the immunohematological systems of species existing along with us.

Blood Groups of Primates, pages 239–254
© 1983 Alan R. Liss, Inc., 150 Fifth Avenue, New York, NY 10011

Bibliography

1. Alepa FP: Antigenic factors characteristic of human immunoglobulin G detected in the sera of non-human primates. Primates in Med 1:1–9, 1968.

1a. Allan FH, Corcoran AP, Ellis FR: Some new observations on the MN system. Vox Sang 6:224–231, 1960.

2. Allen FH, Diamond LK: Erythroblastosis fetalis, including exchange transfusion techniques. Boston: Little, Brown, 1958.

3. Ayala F: Les mécanismes de l'évolution. In l'Evolution—Pour la Science, Vol 13, 1978.

4. Balner H, Dersjant H, van Rood JJ: Leukocyte antigens and histocompatibility in monkeys. Ann NY Acad Sci 129:541, 1966.

4a. Balner H, Vreeswijk van W, Roger HJ, D'Amaro J: The major histocompatibility complex of chimpanzees: Identification of several new antigens controlled by the A and B loci of ChLA. Tissue Antigens 12:1–18, 1978.

5. Balzamo E, Bert J, Menini C, Naquet R: Excessive light sensitivity in *Papio papio:* its variations with age, sex and geographic origin. Epilepsia 16:269–276, 1975.

6. Bernard J, Ruffie J: "Hématologie géographique," Vol 1: "Ecologie humaine, caractères Héréditaires du sang." Paris: Masson, 1966; Vol 2: "Variations hématologiques acquises: l'hématologie et l'évolution," Paris: Masson, 1972.

7. Berkeley HK: The impossibility of differentiation between monkey blood and human blood by means of antisera derived from monkeys. Pathology 2:10–110, 1913.

7a. Bernstein F: Ergebnisse einer biostatischen zusammenfassenden Betrachtung über die erblichen Blutstrukturen des Menschen. Klin Wochenschr 3:1495, 1924.

7b. Bernstein F: Zusammenfassende Betrachtungen über erbliche Blutstrukturen des Menschen. Zschr indukt Abstamm-Vererb Lehre 37:237, 1925.

7c. Beutler E, West C: Glucose-6-phosphate dehydrogenase variants in the chimpanzee. Biochem Med 20:364–370, 1978.

8. Bird GWG: A intermediate in Maharastrian blood donors. Vox Sang 9:629–630, 1964.

9. Blumenfeld OO, Adamany AM: Structural polymorphism within the amino-terminal region of MM, NN and MN glycoproteins (glycophorins) of the human erythrocyte membrane. Proc Natl Acad Sci USA 75:2727–2731, 1978.

9a. Blumenfeld OO, Adamany AM, Puglia KV, Socha WW: The chimpanzee M blood group antigen is a variant of the human M-N glycoproteins. Biochem Genet 21:333–348, 1983.

10. Bogden AE, Gray JH: Isoimmune responses in the rhesus monkey. Hemagglutination, leukoagglutination and leukotoxicity. Books Air Force Base, Texas: USAF School of Aerospace Medicine, Aerospace Medical Division (AFSC), Publication No. SAM-TR-67-14, February 1967.

11. Bogden AE, Gray JH, Brule M: Population frequency of erythrocyte antigens in the rhesus monkeys. Lab Anim Sci 24:910–913, 1974.

12. Boyd WC: Blood groups. Tabul Biol Hague 17ii:113–240, 1939.

13. Brett FL, Jolly CJ, Socha WW, Wiener AS: Human-like ABO blood groups in wild Ethiopian baboons. Yearb Phys Anthropol 20:276–280, 1976.

14. Bridges CB: Salivary chromosome maps. Sci Hered 26:60–64, 1935.

15. Bryant NJ: "An Introduction to Immunohematology." Philadelphia, London, Toronto: W.B. Saunders, 1976.

16. Buchbinder L: The blood grouping of macacus rhesus, including comparative studies of the antigenic structure of the erythrocytes of man and macacus rhesus. J Immunol 25:33–59, 1933.

17. Cartron JP: Biosynthèse des antigènes de groupes sanguins humains. In "XVIIe Cong Soc Intern Hématol Paris, July 23–29, 1978." Paris: Edition Arnette, 1979.

18. Castro O, Socha WW, Moor-Jankowski J: Survival of human erythrocytes in primate animals. J Med Primatol 10:55–60, 1981.

19. Castro O, Socha WW, Moor-Jankowski J: Human sickle erythrocytes: survival in chimpanzees. J Med Primatol 11:119–125, 1982.

20. Cavalli-Sforza L, Bodmer WF: "The Genetics of Human Populations." San Francisco: W.H. Freeman, 1971.

21. Cavalli-Sforza LL, Feldman MW: "Cultural Transmission and Evolution." Monographs in Population Biology. Princeton, NJ: Princeton University Press, 1981.

22. Chakraborty R, Ferrell RE, Schull WJ: Paternity exclusion in primates: two strategies. Am J Phys Anthropol 50:367–372, 1979.

23. Cleghorn F: A memorandum on the Miltenberger blood group. Vox Sang 11:219–222, 1966.

23a. Cleve H, Patuschnick W: Different phenotypes of the group-specific component (Gc) in chimpanzee. Human Genet 50:217–220, 1979.

24. Dahr P: Zeitschrift fur Rassenphysiologie 8:145, 1936 (quoted by Wiener).

24a. Dahr W: Serology, genetics and chemistry of the MNSs blood groups system. Blood Transfus Immunohaematol 24:85–95, 1981.

24b. DeCastello A von, Sturli A: Ueber die Isoagglutinine im Serum gesunder und kranker Menschen. München med Wchschr 49:1090–1095, 1902.

25. Diamond BA, Yelton DE, Scharff MD: Monoclonal antibodies. A new technology for producing serologic reagents. N Engl J Med 304:1344–1349, 1981.

26. Diczfalusy E, Standley CC: "The Use of Nonhuman Primates in Research on Human Reproduction." Stockholm: Karolinska Institute, WHO Research and Training Center on Human Reproduction, 1972.

27. Downing HJ, Benimadho S, Bolstridge MC, Klomfass HJ: The ABO blood groups in vervet monkeys (Cercopithecus pygerythrus F. Cuvier). J Med Primatol 2:290–295, 1973.

28. Downing HJ, Burgers LE, Getliffe FM: A-B-O blood groups of two subspecies of chacma baboons (Papio ursinus) in South Africa. J Med Primatol 4:103–107, 1975.

29. Ducos J, Ruffié J: Contribution à l'étude des différences existant entre les groupes A$_1$ et A$_2$. Montpellier Méd 9:41–42, 1059, 1954.

30. Duggleby CR, Blystad C, Stone WH: Immunogenetic studies of rhesus monkeys. II. The H, I, J, K and L blood group systems. Vox Sang 20:124–136, 1971.

31. Duggleby CR, Stone WH: Immunogenetic studies of rhesus monkeys. I. The G blood group system. Vox Sang 20:109–123, 1971.

32. Dujarric de la Rivière R, Eyquem A: "Les groupes sanguins chez les animaux." Paris: Flammarion, 1953.

32a. von Dungern E, Hirszfeld L: Ueber Vererbung gruppenspezifischer Strukturen des Blutes (II). Zschr Immunitätsf 6:284–285, 1910.

33. von Dungern E, Hirszfeld L: Ueber Vererbung gruppenspezifische Strukturen des Blutes (III). Zschr Immunitätsf 8:526–530, 1911.

34. Dunsford I, Ikin EW, Mourant AE: A human blood group gene intermediate between M and N. Nature 172:688–689, 1953.
35. Dutrillaux B: Nature et origine des chromosomes humains. Annales Génét. l'Expansion scient., Paris, 1975.
36. Dutrillaux B: Chromosomal evolution in primates: tentative phylogeny from microcebus marinus (prosimian) to man. Hum Genet 48:251–314, 1979.
37. Dutrillaux B, Couturier J, Vieges-Pequignot E: Chromosomal evolution in primates. Chromosomes Today 7:176, 1981.
38. Edwards RH: The G and H rhesus monkey blood group system. J Hered 62:79–86, 1971.
39. Edwards RH: The I blood group system of rhesus monkeys. J Hered 62:142–148, 1971.
40. Edwards RH: Rhesus monkey blood group systems J, K, L, M, N and O. J Hered 6:149–156, 1971.
41. Edwards RH: Letter to the editor. Vox Sang 24:561–562, 1973.
41a. Ehrlich P, Morgenroth J: Ueber Haemolysine. Berlin Klin Woch 37:453–458, 1900.
42. Elliot M, Bossom E, Dupuy ME, Masouredis SP: Effect of ionic strength on the serologic behavior of red cell isoantibodies. Vox Sang 9:396–401, 1964.
42a. Epstein AA, Ottenberg R: Simple method of performing serum reactions. Proc NY Path Soc. 8:117, 1908.
43. Erskine AG: The technic of blood grouping. In Frankel S, Reitman S, Sonnenwirth AC (eds): "Gradwohl's Clinical Laboratory Methods and Diagnosis." Saint Louis: C.V. Mosby, 1970, pp 720–751.
44. Erskine AG, Socha WW: "The Principles and Practice of Blood Grouping." 2nd Ed. Saint Louis: C.V. Mosby, 1978.
45. Fischer W, Klinkhart G: Ueber Hämmagglutination und Hämolyse bei Macacus cynomolgus und Simia rhesus. Z Immunitätsforsch 15:513–526, 1932.
46. Foran RF: Blood groups in subhuman primates: a review of the literature with comments. In Pickerring ED (ed): "Proceedings of a Conference on Research with Primates." Beaverton, Oregon: Tektronix Foundation, 1963, pp 75–79.
47. Fortner JG, Beattie EJ, Hei Shiu Man, Howland WS, Sherlock P, Moor-Jankowski J, Wiener AS: The treatment of hepatic coma in man by cross-circulation with baboon. In Goldsmith EI, Moor-Jankowski J (eds): "Medical Primatology, 1970." Basel, New York: Karger, 1971, pp 62–68.
48. Friedenreich V: "The Thomsen Hemagglutination Phenomenon." Copenhagen: Levin and Munksgaard, 1930.
49. Froehlich JW, Socha WW, Wiener AS, Moor-Jankowski J, Thorington RW: Blood groups of the mantled howler monkey (*Alouatta palliata*). J Med Primatol 6:219–231, 1972.
49a. Furthmayr H: Structural comparison of glycophorins and immunochemical analysis of genetic variants. Nature (London) 271:519–524, 1978.
49b. Gahmberg C, Jokinen M, Karhi K, Ulmanen I, Kaarianinen L, Anderson L: Biosynthesis of the major human red cell sialoglycoprotein. Blood Transfus Immunohaematol 24:53–73, 1981.
50. Gedde-Pahl T, Grimstad AL, Gundersen S, Vogt E: A probable crossing over or mutation in the MNS blood system. Acta Genet 17:193–210, 1967.
51. Gengozian N: Human A- and B-like antigens on red cells of marmosets. Proc Soc Exp Biol Med 117:858–865, 1966.
52. Gengozian N: Inter- and Intrasubspecies red cell immunizations in marmosets, *Saguinus fuscicollis* ssp. J Med Primatol 1:172–192, 1972.

53. Gengozian N: A blood factor in marmosets, *Saguinus fuscicollis*. Its detection, mode of inheritance, and species specificity. J Med Primatol 1:272–286, 1972.

54. Gengozian N, Lushbaugh CC, Humason GL, and Kniseley RM: 'Erythroblastosis foetalis' in the primate *Tamarinus nigricollis*. Nature (London) 209:731–732, 1966.

55. Gengozian N, Patton ML: Identification of three blood factors in the marmoset, *Saguinus fuscicollis spp*. In Goldsmith EI, Moor-Jankowski J (eds): "Medical Primatology, 1972," Part I. Basel: Karger, 1972, pp 349–360.

56. Gibbs MS, Akeroyd JH, Zapf JJ: Quantitative subgroups of the B antigen in man and their occurrence in three racial groups. Nature 192:1196, 1961.

57. Giorno R: Model for the Rh blood group system, based on discontinuous gene structure. Vox Sang 41:102–109, 1981.

58. Goldsmith EI: Unpublished data.

59. Goldsmith EI, Moor-Jankowski J, Wiener AS, Allen FH, Hirsch R: Exchange transfusion of nonhuman primates with human blood. A program for preparation of cross-circulation partners in hepatic failure. In Goldsmith EI, Moor-Jankowski J (eds): "Medical Primatology, 1970." Basel, New York: Karger, 1971, pp 80–89.

59a. Goodman M: Evolution of the catarrhine primates at the macromolecular level. Primates in Med 1:10–26, 1968.

60. Gorer PA, Mikulska ZB: The antibody response to tumor inoculation. Improved methods of antibody detection. Cancer Res 14:651–670, 1954.

61. Grouchy (de) J: "De la Naissance des Espèces aux Aberrations de la Vie." Paris: Robert Laffont, 1978.

62. Grove-Rasmussen M, Shaw RW, Marceau E: Hemolytic transfusion reaction in group-A patient receiving group-O blood containing immune anti-A antibodies in high titer. Am J Clin Pathol 23:828–835, 1953.

62a. Groves CP: Phylogeny and classification of primates. In T-W-Fiennes RN (ed.): "Pathology of Simian Primates," Part I, pp 11–57. Basel: Karger, 1972.

63. Hakomori S, Watanabe K, Laine RA: Glycophospholipids with blood group A, H, and I activity. Their status in group A_1 and A_2 erythrocytes and their changes associated with ontogeny and oncogeny. In "Human Blood Groups. 5th Int Convoc Immunol, Buffalo, N.Y., 1976." Basel: Karger, 1977, pp 150–163.

63a. Haynes BF, Dowell BL, Hensley LL, Gore I, Metzgar RS: Human T cell antigen expression by primate T cells. Science 215:298–300, 1982.

64. Hirose Y, Balner H: Red cell iso-antigens of rhesus monkeys. Blood 34:661–681, 1969.

65. Hirsch W, Moores P, Sanger R, Race R: Notes on some reactions of human anti-M and anti-N sera. Br J Hematol 3:134–142, 1957.

65a. Hirszfeld L, Hirszfeld H: Serological differences between the blood of different races. Lancet ii:675–679, 1919.

65b. Hirszfeld L, Przesmycki F: Badania nad aglutinacja normalna. IV. O izoaglutynacji u koni (Study on normal agglutination. IV. Iso-agglutination in horses). Przegl Epidemiol 1:577–578, 1921.

66. Hoffstetter R: Phylogeny and geographical deployment of the primates. J Hum Evol 3:327, 1974.

67. Hume DM, Gayle WE, Williams GM: Cross-circulation of patients in hepatic coma with baboon partners having human blood. Surg Gynecol Obstet 128:495–501, 1969.

67a. Ikin AW, Mourant AE: A rare blood group antigen occurring in negroes: Br Med J 1:456–457, 1951.

68. Irwin MR: Genetics and Immunology. In Dunn LC (ed): "Genetics in the 20th Century." New York: Macmillan, 1951, 173–219.

69. Irwin MR: Blood grouping and its utilization in animal breeding. Proceedings VIIth Int Cong Anim Husbandry and Production, Suppl 2:1–41, 1956.

70. Iseki S: ABH blood groups substances in living organisms. "Human Blood Groups, 5th Int Convoc Immunol, Buffalo NY, 1976." Basel: Karger, 1977, pp 126–133.

70a. Jack JA, Tippett P, Woodes J, Singer R, Race R: M₁: A subdivision of the human blood group antigen M. Nature (London) 186:642, 1960.

71. Jolly CJ, Turner TR, Socha WW, Wiener AS: Human-type A-B-O blood group antigens in Ethiopian vervet monkeys *(Cercopithecus aethiops)* in the wild. J Med Primatol 6:54–57, 1977.

72. Kariher DH, Miller DL: Evidence of maternal Rh sensitization without evidence of hemolytic disease of the newborn. Am J Med Sci 212:327–330, 1946.

73. Kellner A, Hedal EF: Experimental erythroblastosis fetalis. Am J Pathol 28:539, 1952.

74. Kellner A, Hedal EF: Experimental erythroblastosis fetalis in rabbits. I. Characterization of a pair of allelic blood group factors and their specific immune isoantibodies. J Exp Med 97:33–49, 1953.

75. Kellner A, Hedal EF: Experimental erythroblastosis fetalis in rabbits. II. The passage of blood group antigens and their specific isoantibodies across the placenta. J Exp Med 97:57–60, 1953.

76. Kennet RH, McKearn TJ, Bechtol KB (eds): "Monoclonal Antibodies: Hybridomas; A New Dimension in Biological Analyses." New York: Plenum, 1980.

77. Kohler G, Milstein C: Continuous cultures of fused cells secreting antibody of predefined specificity. Nature (London) 256:495–497, 1975.

78. Kimura M: Evolutionary rate at the molecular level. Nature 217:624, 1963.

79. Kimura M: Genetic variability maintained in a finite population due to mutational production of neutral and nearly neutral isoalleles. Genet Res 11:247, 1968.

80. Kimura M: Gene pool of higher organisms as a product of evolution. Cold Spring Harbor Symp Quant Biol 38:515.

81. Kimura M: La théorie neutraliste de l'évolution moléculaire. Pour la Science (Scientific American) 27:48, 1980.

82. Kimura M, Ohta T: Mutation and evolution at the molecular level. Genetics Suppl 73:20, 1973.

83. Lamotte M: "Initiation aux Méthodes Statistiques en Biologie." Paris: Masson, 1957.

84. Landois W: "Die Transfusion des Blutes." Leipzig 188. Cited in Keynes "Blood transfusion." London: Wright, 1949.

85. Landsteiner K: Uber Agglutinationserscheinungen normalen menchschlichen Blutes. Klin Wochenschr 14:95–95, 1901.

86. Landsteiner K, Levine P: On the cold agglutinins in human serum. J Immunol 12:441–460, 1926.

87. Landsteiner K, Levine P: A new agglutinable factor differentiating individual human bloods. Proc Soc Exp Biol Med 24:600, 1927.

88. Landsteiner K, Levine P: Further observations on individual differences of human blood, 1927. Proc Soc Exp Biol Med 24:941, 1927.

89. Landsteiner K, Levine P: On individual differences in human blood. J Exp Med 47:757–775, 1928.

90. Landsteiner K, Levine P: On the inheritance of agglutinogens of human blood demonstrable by immune agglutinins. J Exp Med 48:731–749, 1928.

91. Landsteiner K, Levine P: On isoagglutinin reactions of human blood other than those defining the blood groups. J Immunol 17:1–28, 1929.

91a. Landsteiner K, Levine P: Differentiation of a type of human blood by means of normal animal serum. J Immunol 18:87–94, 1930.

92. Landsteiner K, Levine P: Immunization of chimpanzees with human blood. J Immunol 22:397–400, 1932.

93. Landsteiner K, Miller CP: Serological studies on the blood of the primates. I. The differentiation of human and anthropoid blood. J Exp Med 42:841–852, 1925.

94. Landsteiner K, Miller CP: Serological studies on the blood of the primates. II. Distribution of serological factors related to human isoagglutinogens in the blood of lower monkeys. J Exp Med 42:863–872, 1925.

95. Landsteiner K, Miller CP Jr: Serological studies on the blood of priamtes. III. Distribution of serological factors related to human isoagglutinogens in the blood of lower monkeys. J Exp Med 42:863–870, 1925.

96. Landsteiner K, Miller CP: Serological studies on the blood groups of anthropoid apes. J Exp Med 42:853–862, 1925.

96a. Landsteiner K, Strutton WR, Chase MW: An agglutination reaction observed with some human blood, chiefly among Negroes. J Immunol 27:469–472, 1934.

97. Landsteiner K, Wiener AS: On the presence of M agglutinogens in the blood of monkeys. J Immunol 33:19–23, 1937.

98. Landsteiner K, Wiener AS: An agglutinable factor in human blood recognized by immune sera for blood. Proc Soc Exp Biol Med 43:223, 1940.

99. Landsteiner K, Witt DW: Observations on the human blood groups. Irregular reactions. Isoagglutinins in sera of group IV. The factor A_1. J Immunol 11:221–247, 1926.

100. LaSalle M: Erythrocyte isoantigens of the genus Macaca. In Hofer HO (ed): "Proc 2nd Internatl Congr Primatol, Atlanta, Georgia, 1968, Vol 3. Basel: Karger, 1969, pp 120–128.

101. LaSalle M: Erythrocyte isoantigens in rhesus monkeys: Factors IIB and IIC. Am J Vet Res 32:445–455, 1971.

102. LaSalle M: Interrelationships of blood groups between rhesus macaques *(Macaca mulatta)* and Celebes apes *(Macaca niger)*. Folia Primatol 70:95–100, 1973.

103. LaSalle M, DeLannoy CW: Immunologic response to blood transfusion in subhuman primates. Am J Vet Res 30:428–433, 1969.

104. LaSalle M, Frisch AW: Species specific blood groups in *Cynopithecus niger* (Celebes Ape). Proc Soc Exp Biol Med 118:940–943, 1965.

105. Levine P, Birnbaum L, Katzin EM, Vogel P: The role of isoimmunization in the pathogenesis of erythroblastosis fetalis. Am J Obstet Gynecol 42:925–941, 1941.

106. Levine P, Kumichel AB, Wigod M, Kock E: New blood factor, s, allelic to S. Proc Soc Exp Biol Med 78:218, 1951.

107. Levine P, Ottensooser F, Celano MJ, Pollitzer W: On reactions of plant anti-N with red cells of chimpanzees and other animals. Am J Phys Anthropol 13:29–36, 1955.

108. Levine P, Stetson RE: An unusual case of intragroup agglutination. J Am Med Assoc 113:126–127, 1939.

108a. Levine P, Stock AH, Kumichel AB, Bronikowsky N: A new human blood factor of rare incidence in the general population. Proc Soc Exp Biol Med 77:402–403, 1951.

109. Lewin R: Evolutionary history written in globin genes. Science 214:426–429, 1981.

110. Lewontin R, Hubby JL: A molecular approach to the study of genetic heterozygosity in natural populations. Genetics II 54:595, 1966.

110a. Lisowska E: Biochemistry of M and N blood group specificities. Blood Transfus Immunohaematol 24:75–84, 1981.

111. Lisowska E, Kordowicz M: Immunochemical properties of M and N blood groups antigens and their degradation products. In "Human Blood Groups, 5th Int Convoc Immunol, Buffalo, NY, 1976." Basel: Karger, 1977, pp 188–196.

112. Van Loghem E, Socha WW, de Lange G: Production of antibodies to human immunoglobulins in a gibbon. Vox Sang 37:262–267, 1979.

113. Lopas H, Birndorf NI: Haemolysis and intravascular coagulation due to incompatible red cell transfusion in isoimmunized rhesus monkeys. Br J Haematol 21:399–411, 1971.

114. Lucotte G: Similitudes génétiques comparées entre différentes espèces, *Papio papio, P. anubis, P. cynocephalus* et *P. hamadryas,* basées sur les données de polymorphisme des enzymes érythrocytaires. Biochem Systemat 7:245–252, 1979.

115. Lucotte G: Distances électrophorétiques entre les différentes espèces de singes du groupe des Mangabeys. Ann Génét 22:85–87, 1979.

116. Lucotte G: Polymorphisme électrophorétique des protéines et enzymes sériques et érythrocytaires chez le chimpanzé. Hum Genet 54:97–102, 1980.

117. Lucotte G, Guillon R: Polymorphisme électrophorétique des protéines et enzymes sériques chez le babouin de Guinée Papio papio. Biochem Systemat Ecol 7(3):239–244, 1979.

118. Majdic O, Knapp W, Vetterlein M, Mayr WR, Speiser P: Hybridomas secreting monoclonal antibodies to human group A erythrocytes. Immunobiology 156:226–227, 1979.

118a. Marsh WL: Present status of the Duffy blood group system. CRC Crit Rev Clin Lab Sci, pp 387–412, March, 1975.

119. Masouredis SP, Dupuy ME, Elliot M: Distribution of the Rh (D) antigen in the red cells of non-human primates. J Immunol 98:8–16, 1967.

120. McDermin EM, Vos GH, Downing HJ: Blood groups, red cell enzymes, and serum proteins of baboons and vervets. Folia Primatol 19:312–326, 1973.

120a. Meera Khan P, Balner H: Polymorphic enzymes in rhesus monkeys and chimpanzees. In Goldsmith EI, Moor-Jankowski J (eds): "Medical Primatology, 1972," Part I. Basel: Karger, 1972, pp 363–371.

121. Milstein C: Monoclonal antibodies from hybrid myelomas. (The Wellcome Foundation Lecture, 1980.) Proc Roy Soc Lond B 21:393–412, 1980.

122. Mohn JF, Cunninghan RK, Bates JF: Qualitative distinction between subgroups A$_1$ and A$_2$. In "Human Blood Groups, 5th Int Convoc Immunol, Buffalo, NY, 1976." Basel: Karger, 1977, pp 316–325.

123. Moon GJ, Wiener AS: A new source of anti-N lectin: Leaves of the Korean *vicia unijuga.* Vox Sang 26:167–169, 1974.

124. Moor-Jankowski J, Socha W: Blood groups of macaques: a comparative study. J Med Primatol 7:136–145, 1978.

125. Moor-Jankowski J, Socha W: Blood groups of Old World monkeys. Evolutionary and taxonomic implications. J Hum Evol 8:44–45, 1979.

126. Moor-Jankowski J, Socha WW, Wiener AS, Plonski H: Chimpanzee simian-type blood groups: Reproducibility of formerly described antisera and demonstration of new blood groups O^c and P^c. Folia Primatol 28:216–230, 1977.

127. Moor-Jankowski J, Wiener AS: Simian type blood factors in non human primates. Nature (London) 205:369–371, 1965.

128. Moor-Jankowski J, Wiener AS: Blood groups of nonhuman primates; summary of the currently available information. Primates Med 1:95–99, 1968.

129. Moor-Jankowski J, Wiener AS: Blood group antigens in primate animals and their relation to human blood groups. Primates Med 3:64–77, 1969.

130. Moor-Jankowski J, Wiener AS: Red cell antigens of primates. In RNT-W. Fiennes (ed): "Pathology of Simian Primates." Basel: Karger, 1972, Part 1, pp 270–317.

131. Moor-Jankowski J, Wiener AS, Gordon EB: Blood groups of apes and monkeys. III. The MN blood factors of apes. Folia Primatol 2:129–148, 1964.

132. Moor-Jankowski J, Wiener AS, Gordon EB: Blood group antigens and cross-reacting antibodies in primates, including man. III. Heterophile-like behavior of the blood factor I. Exp Med Surg 22:308–315, 1964.

133. Moor-Jankowski J, Wiener AS, Gordon EB: Simian blood groups. Two "new" blood factors of gibbon blood, A^g and B^g. Transfusion 5:235–239, 1965.

134. Moor-Jankowski J, Wiener AS, Gordon EB, Davis JH: Blood groups of monkeys demonstrated with isoimmune rhesus monkey sera. Int Arch Allergy Appl Immunol 30:373–377, 1967.

135. Moor-Jankowski J, Wiener AS, Gordon EB, Davis JH: Blood groups of baboons, demonstrated with isoimmune sera. Nature (London) 214:181–182, 1967.

136. Moor-Jankowski J, Wiener AS, Gordon EB, Guthrie CB: Simian blood groups. A new blood factor A^{ba} of Celebes black apes demonstrated with rabbit antisera. Folia Primatol 3:245–248, 1965.

137. Moor-Jankowski J, Wiener AS, Rogers CM: Blood groups of chimpanzees: demonstrated with isoimmune serums. Science 14:1441–1443, 1964.

138. Moor-Jankowski J, Wiener AS, Socha WW: A new taxonomic tool. II. Serological differences between baboons and geladas demonstrated by cross-immunization. Folia Primatol 22:59–71, 1974.

139. Moor-Jankowski J, Wiener AS, Socha WW, Gordon EB, Davis JH: A new taxonomic tool: Serological reactions in cross-immunized baboons. J Med Primatol 2:71–84, 1973.

140. Moor-Jankowski J, Wiener AS, Socha WW, Gordon EB, Kaczera Z: Blood group homologues in orangutans and gorillas of the human Rh-Hr and chimpanzee C-E-F system. Folia Primatol 19:360–367, 1973.

141. Moor-Jankowski J, Wiener AS, Socha WW, Gordon EB, Mortelmans J, Sedgwick CJ: Blood groups of pygmy chimpanzees *(Pan paniscus):* human-type and simian-type. J Med Primatol 4:262–267, 1975.

142. Moor-Jankowski J, Wiener AS, Socha WW, Valerio DA: Blood groups of crab-eating macaques *(Macaca fascicularis)* demonstrated by isoimmune rhesus monkey *(Macaca mulatta)* sera. J Med Primatol 6:76–86, 1977.

143. Moore HC, Mollison PL: Use of a low-ionic strength medium in manual tests for antibody detection. Transfusion 16:167–169, 1974.

144. More KN, Banejee K: Homologues of the human A, B, O & M, N blood groups in *Macaca radiata* (Geoffroy) & *Presbitis entellus* (Dufresne) & the reactions of their erythrocytes with lectins. Indian J Exp Biol 17:1330–1332, 1979.

145. Mourant AE: "The Distribution of the Human Blood Groups." Oxford: Blackwell Scientific Publications, 1954.

146. Mourant AE, Kopec A, Domaniewska-Sobczak K: "The Distribution of the Human Blood Groups and Other Polymorphisms," 2nd ed. Oxford: Oxford University Press, 1976.

147. Mourant AE, Kopec AC, Domaniewska-Sobczak K: "Blood Groups and Diseases." Oxford: Oxford University Press, 1978.

148. Mourant AE, Race RR: The Rh system in the chimpanzee. Science 104:277, 1946.

149. Mourant AE, Ruffié J: Seminar at the Collège de France, February 22, 1979.

150. Muchmore E, Socha WW: Blood transfusion therapy for leukemic chimpanzee. Lab Primate Newsl 15:13–15, 1976.

151. Murphy GP, Mirand EA, Groenewald JH: Erythropoietin release in baboon renal allografts. In Goldsmith EI, Moor-Jankowski J (eds): "Medical Primatology, 1971." Basel: Karger, 1971, pp 103–118.

152. Napier JR, Napier PH: "A Handbook of Living Primates." 3rd Ed. London, New York: Academic, 1970.

153. Neimann-Sorensen A: "Blood Groups of Cattle: Immunogenetic Studies on Danish Cattle Breeds." Copenhagen: A/S Carl F. Mortensen, 1958, pp 29–51.

154. Ohno S: "Evolution by Gene Duplication." Berlin, Heidelberg, New York: Springer-Verlag, 1970.

154a. Oriol R: Interaction of ABO, Hh, Secretor and Lewis systems. Blood Transfus Immunohaematol 23:517–526, 1980.

154b. Oriol R, Danilovs J, Hawkins BR: A new genetic model proposing that the Se gene is a structural gene closely linked to the H gene. Am J Hum Genet 33:421–431, 1981.

154c. Oriol R, Le Pendu J, Lemieux RU: Chemistry and genetics of H-h and Se-se blood group systems. Congr Intl Soc Haematol, Budapest, 1982 (Abstract).

154d. Oriol R, Le Pendu J, Sparkes RS, Crist M, Gale RP, Terasaki P, Bernoco M: Insights into the expression of ANH and Lewis antigens through human bone marrow transplantation. Am J Hum Genet 33:551–560, 1981.

155. Ottensooser F, Silberschmidt K: Haemagglutinin anti-N in plant seeds. Nature (London) 172:914, 1953.

156. Owen RD, Anderson DR: Blood groups in rhesus monkeys. Ann NY Acad Sci 97:4–8, 1962.

156a. Palatnik M, Rowe A: Duffy and Duffy-related human antigens in primates. J Hum Evol (in press).

157. Pardoe GI, Uhlenbruck G, Reifenberg U: Structural studies of red cells: immunochemistry of the MN and related antigens and the mqx viral receptors. Med Lab Technol 28:255, 1971.

158. Potter EL: Rh: Its relation to congenital hemolytic anemia and to intragroup transfusion reactions. Chicago: Year Book Publishers, 1974.

158a. Race R: The Rh genotype and Fisher's theory: Blood 3:27–42, 1948.

159. Race R, Sanger R: "Blood Groups in Man," 6th ed. London: Blackwell Scientific Publications, 1975, p 109.

160. Reemtsma K, McCraken BH, Schlegel JU, Pearl MA, Dewitt CW, Hewitt RL, Creech O Jr: Renal heterotransplantation in man. J Am Med Assoc 187:691–702, 1964.

161. Reinberg A, Ghata J: "Les rhythmes biologiques." Que sais-je. Paris: PUF, 1957.

161a. Report of the Ad Hoc Task Force to Develop a National Chimpanzee Breeding Program of the Interagency Primate Steering Committee, Tanglewood, North Carolina, June 2–3, 1980. Bethesda, MD: U.S. Department of Health and Human Services, Public Health Service, National Institutes of Health, 1982.

162. Roberts GF: Comparative aspects of haemolytical disease. London: Heinemann, 1975.

163. Ropars C, Cartron JP: Aspects bio-chimiques des antigènes de groupes sanguins. Rev Franç Transfus Immuno-Hématol 1:193–206, 1976.

164. Rosenfield R, Allen A, Rubinstein P: Genetic model for the Rh blood group system. Proc Natl Acad Sci USA 70:1303–1307, 1973.

165. Rosenfield RE, Haber Gladys V, Rutte S, Rachel B: A negro family revealing Hunter-Henshaw information, and independence of the genes for Js and Lewis. Am J Hum Genet 12:143–146, 1960.

166. Rowe AW, Allen FH: Freezing of blood droplets in liquid nitrogen for use in blood group studies. Transfusion 5:379–382, 1965.

167. Rowe AW, Davis JE: Erythrocyte survival in chimpanzees, gibbons and baboons. J Med Primatol 1:86–89, 1972.

168. Rowe AW, Davis JH, Moor-Jankowski J: Preservation of red cells from the nonhuman primates. Primates Med 7:117–130, 1972.

169. Rowe AW, Eyster E, Kellner A: Liquid nitrogen preservation of red blood cells for transfusion. A low glycerol rapid freeze procedure. Cryobiology 5:119–122, 1968.

170. Rowe AW, Miles W, Vermeulen J, Moor-Jankowski J: An automated apparatus for droplet freezing of blood in liquid nitrogen. Cryobiology 3:385–390, 1967.

171. Rubinstein P, Puza A, Vlahovic S, Ferrebee JW: Isohemagglutination in dogs. The dextran method. Folia Biol 10:36–42, 1964.

171a. Ruffié J: "Les Groupes Sanguins chez l'Homme." Paris: Masson, 1956.

172. Ruffié J: "Hémotypologie, Evolution du Groupe Humain." Paris, Ermann, 1966.

173. Ruffié J: Les frontières chromosomiques de l'hominisation. CR Acad Sci 276(11):1709–1711, 1973.

174. Ruffié J: Les données de l'immunogénétique et de la cytogénétique et le monophylétisme humain. Anthropologie 75:57–84, 1971.

175. Ruffié J: Les données de l'immunogénétique et le processus de spéciation chez les primates. CR Acad Sci Paris D 276:2101–2104, 1973.

176. Ruffié J: La signification anthropologique des antigènes publics et des antigènes privés. CR Acad Sci 277:24, 2715, 1975.

177. Ruffié J: Blood groups and primate evolution. XVIth Congress Soc Int Hématol, Kyoto, 1976.

178. Ruffié J: "De la Biologie à la Culture." Paris: Flammarion, 1976.

179. Ruffié J: Spéciation populationnelle ou spéciation chromosomique? Annales de génétique. Sem Hop Paris 55:286–291, 1979.

180. Ruffié J: Sur les paratypes: antigènes communs à certains groupes de primates. CR Acad Sci Paris 27:2437, 1979.

181. Ruffié J: Spéciation chromosomique ou spéciation populationnelle. Sem Hop Paris 55(5–6):286–291, 1979.

182. Ruffié J: "Populations et Sociétés." Paris: Flammarion, 1982.

183. Ruffié J, Bernard J: Hématologie géographique et dynamique des populations. Nouv Rev Fn Hématol 21:321–246, 1979.

184. Ruffié J, Bouloux C: Sur la présence d'une substance inhibitrice de l'anticorps humain anti-i dans plusieurs espèces de singes de l'ancien et du nouveau monde. CR Acad Sci Paris 272:1709–1712, 1971.

185. Ruffié J, Bouloux C, Ruffié P: Sur les anticorps naturels anti-primates du serum humain. CR Acad Sci Paris 269:252–264, 1969.

186. Ruffié J, Marty Y: Heteragglutinins in human sera specific for erythrocytes of non human primates. J Med Primatol 1(4):211–219, 1972.

187. Ruffié J, Schoenwetter D: Sur la variabilité des anticorps anti-singes dans le serum humain AB. CR Acad Sci Paris 272:159–162, 1971.

187a. Ruffié J, Socha WW: Les groupes sanguins érythrocytaires des primates non-hominiens. Nouv Rev Fr Hématol 22:147–209, 1980.

188. Sacks SH, Lennox E: Monoclonal anti-B as a new blood typing reagent. Vox Sang 40:99–10, 1981.

189. Salmon D: Aide aux diagnostiques de paternité à partir des marqueurs du polymorphisme génétique. Th Biol Hum: Paris 6 UER Pitié Salpêtrière: 1977; 3.

190. Sanger R: "Blood Groups in Human Genetics." Witebsky Memorial lecture. In "Human Blood Groups 5th Int Convoc Immunol, Buffalo, New York, 1976." Basel: Karger, 1977, pp 1–16.

191. Seigler HR, Patterson J, Metzgar RS, Zwiren GT, MacDonnel RC Jr, Behrens BL, Corrigan JJ: Cross-circulation between a patient in hepatic coma and a chimpanzee. In Goldsmith EI, Moor-Jankowski J (eds): "Medical Primatology, 1970." Basel: Karger, 1971, pp 69–79.

191a. Simpson GG: The principles of classification and a classification of mammals. Bull Amer Mus Nat Hist 85:1–350, 1945.

192. Shapiro M: Inheritance of the Henshaw blood factor. J Forensic Med 3:152–160, 1956.

193. Slapak M, Chir M, Baddeley M, Wexler M, Savaris C, Sise H, Garcia S, Giouard M, McDermott WV Jr: Extended preservation of the primate liver by simple cooling and orthotopic autotransplantation. In Goldsmith EI, Moor-Jankowski J (eds): "Medical Primatology, 1970." Basel: Karger, 1971, pp 119–129.

194. Socha WW: "Problems of Serological Differentiation of the Population." Warsaw: State Medical Publications, 1966.

195. Socha WW: Blood groups as genetic markers in chimpanzees: Their importance for the National Chimpanzee Breeding Program. Am J Primatol 1:3–13, 1981.

195a. Socha WW: Blood groups of pygmy and common chimpanzees: a comparative study. In Susman RL (ed): "The Pygmy Chimpanzee: Evolutionary Morphology and Behavior." New York: Plenum (in press).

196. Socha WW, Gordon EB, Saltzman M, Wiener AS: The use of antiglobulin inhibition test for serological studies in primates (seroprimatology). In: "Proc Third Congr Primatol Zurich, 1970." Basel: Karger, 1971, Vol 2, pp 70–80.

197. Socha WW, Lasano SG: The A-B-O blood groups of rhesus monkeys: New observations. J Med Primatol 9:83–87, 1980.

198. Socha WW, Merz E, Moor-Jankowski J: Blood groups of Barbary apes *(Macaca sylvanus)*. Folia Primatol 36:212–225, 1981.

199. Socha WW, Moor-Jankowski J: Rh antibodies produced by an isoimmunized chimpanzee: reciprocal relationship between chimpanzee isoimmune sera and human anti-Rh reagents. Int Arch Allergy Appl Immunol 56:30–38, 1978.

200. Socha WW, Moor-Jankowski J: Blood groups of anthropoid apes and their relationship to human blood groups. J Hum Evol 8:453–465, 1979.

201. Socha WW, Moor-Jankowski J: Serological materno-fetal incompatibility in nonhuman primates. In Ruppenthal RC (ed): "Nursery Care of Nonhuman Primates." New York, London: Plenum, 1979, pp 35–42.

202. Socha WW, Moor-Jankowski J: Chimpanzee R-C-E-F blood group system: A counterpart of the human Rh-Hr blood groups. Folia Primatol 33:172–188, 1980.

203. Socha WW, Moor-Jankowski J, Ruffié J: The graded B^P blood group system of baboons; its relationship with macaque red cell antigens. Folia Primatol (in press).

204. Socha WW, Moor-Jankowski J, Sackett GP: Blood groups of pig-tailed macaques *(Macaca nemestrina)*. Am J Phys Anthropol 48:321–330, 1978.

205. Socha WW, Moor-Jankowski J, Wiener AS, Risser DR, Plonski H: Blood groups of bonnet maques *(Macaca radiata)* with a brief introduction to seroprimatology. Am J Phys Anthropol 45:489–491, 1976.

206. Socha WW, Rowe AW, Lenny LL, Lasano SG, Moor-Jankowski J: Transfusion of incompatible blood in monkeys. Lab Anim Sci 32:48–56, 1982.

207. Socha WW, Ruffié J: Sur la nomenclature du système érythrocytaire D^{rh} des macaques. (in press).

208. Socha WW, Wiener AS, Gordon EB, Moor-Jankowski J: Methodology of primate blood grouping. Transplant Proc 4:107–111, 1972.

209. Socha WW, Wiener AS, Moor-Jankowski J, Jolly CJ: Blood groups of baboons. Population genetics of feral animals. Am J Phys Anthropol 47:435–442, 1977.

210. Socha WW, Wiener AS, Moor-Jankowski J, Scheffrahn W, Wolfson SK Jr: Spontaneously occurring agglutinins in primate sera. Int Arch Allergy Appl Immunol 51:656–670, 1976.

211. Socha WW, Wiener AS, Moor-Jankowski J, Valerio D: The first isoimmune blood group system of rhesus monkeys *(Macaca mulatta):* the graded D^{rh} system. Int Arch Allergy Appl Immunol 52:355–363, 1976.

212. Springer GF: Microbes and higher plants. Their relations to blood groups substance. Proc 10th Cong Int Soc Blood Transf, Stockholm, 1964.

213. Springer GF: Role of human cell surface structure in interactions between man and microbes. Naturwissenschaften 57:162–171, 1970.

214. Springer GF, Desai RR: Human blood group MN and precursor specificities. Structural and biological aspects. Carbohydrat Res 40:183–192, 1975.

215. Springer GF, Desai PR, Yang HJ, Schachter H, Narasinihan S: Interrelations of blood group M and precursor specificities and their signification in human carcinoma. "Hum Blood Groups, 5th Int Convoc Immunol, Buffalo, NY, 1976." Basel: Karger, 1977, pp 179–187.

216. Springer GF, Huprikar V, Tegtmeyer H: Biochemical genetics of human blood group MN specificities and their relation to infectious mononucleosis and oncogenic virus receptors. Z Immunol Forsch 142:99–107, 1971.

217. Startzl TE: "Experience in Renal Transplantation." Philadelphia, London: WB Saunders, 1964.

218. Stevens AR Jr, Finch CA: A dangerous universal donor. Acute renal failure following transfusion of group O blood. Am J Clin Pathol 24:612–620, 1954.

219. Stone WH: Immunogenetics of type-specific antigens in animals. In Greenwall TJ (ed): "Advances in Immunogenetics." Philadelphia: Lippincott, 1967.

220. Stone WH: Letter to the editor. Vox Sang 24:560–561, 1973.

221. Stone WH, Stong R, Blazkovec A, Blystad C, Houser WD: Animal models of mechanisms of erythrocyte destruction: why aren't the erythrocytes of the newborn rhesus monkey *(Macaca mulatta)* destroyed by maternal antibodies. In Sandler SG, Nusbacher J, Schanfield MS (eds): "Immunology of the Erythrocyte." New York: Alan R. Liss, 1980, pp 237–249.

222. Sullivan PT, Blystad C, Stone WH: Immunogenetic studies of rhesus monkeys. IV. Serologic and genetic tests with reagents from Wisconsin and the Netherlands. Transplant Proc 4:117–121, 1972.

223. Sullivan PT, Blystad C, Stone WH: Immunogenetic studies of rhesus monkeys. VII. A simple hemagglutination technique for blood typing. J Immunol Methods 14:31–36, 1977.

224. Sullivan PT, Blystad C, Stone WH: Immunogenetic studies of rhesus monkeys. VIII. A new reagent of the G blood group system and an example of the founder principle. Anim Blood Groups Biochem Genet 8:49–53, 1977.

225. Sullivan PT, Blystad C, Stone WH: Immunogenetic studies of rhesus monkeys. IX. The O, P, Q, R and S blood group systems. Immunogenetics 7:125–130, 1977.

226. Sullivan PT, Blystad C, Stone WH: Immunogenetic studies on the rhesus monkey *(Macaca mulatta):* XI. Use of blood groups in problems of parentage. Lab Anim Sci 27:348–351, 1977.

227. Sullivan P, Duggleby C, Blystad C, Stone WH: Transplacental isoimmunization in rhesus monkey. Abstract Proc Fed Am Soc Exp Biol 31:792, 1972.

228. Szalay FS, Delson E: "Evolutionary History of the Primates." New York: Academic, 1979.

228a. Szymanowski Z, Wachler B: Contribution à l'étude des immunoisoagglutinines du sang du porc. C R Soc Biol 95:932–933, 1926.

229. Terao K, Fujimoto K, Cho F, Honjo Sh: Inheritance mode and distribution of human-type ABO blood groups in the cynomolgus monkey. J Med Primatol 10:72–80, 1981.

229a. Tippett P: Serological study of the inheritance of unusual Rh and other blood group fellow types. Ph.D. Thesis, University of London, 1963.

230. Tobiška J: "Die Phytohämaglutininen." Berlin: Akademie Verlag, 1964, pp 85–87.

230a. Todd C: Cellular individuality in higher animals, with special reference to the individuality of the red blood corpuscle. Proc Roy Soc, ser. B. 106:20–49, 1930.

230b. Todd C: Cellular individuality in higher animals, with special reference to the individuality of the red blood corpuscle. II. Proc Roy Soc, ser. B. 107:197–205, 1930.

230c. Todd C, White RG: On the hemolytic immune isolysins of the ox and their relation to the question of individuality and blood relationship. J Hyg 10:185–195, 1910.

230d. Tsumori A: Newly acquired behavior and social interactions in Japanese monkeys. In "Social Communication Among Primates." Chicago: 1967, pp 207–220.

231. Uhlenbruck G: Possible genetical pathways in the MNSsUu system. Vox Sang 16:200, 1969.

232. Unger MS: Precautions necessary in the selection of a donor for blood transfusion. J Am Med Assoc 76:9–112, 1921.

233. Verbickij MS: The use of hamadryas baboons for the study of immunological aspects of human reproduction. In Diczfalusy E, Standly CC (eds): "The Use of Nonhuman Primates in Research of Human Reproduction." Stockholm: WHO Research and Training Centre on Human Reproduction, Karolinska Inst, 1972, pp 492–505.

234. Voak D, Sacks SH, Alderson T, Takei F, Lennox E, Jarvis J, Milstein C, Darnborough J: Monoclonal anti-A from a hybrid-myeloma: evaluation as a blood grouping reagent. Vox Sang 39:134–140, 1980.

235. Volkova LS, Verbitsky MS, Andreyev AV: Studies on blood groups of lower monkeys in Sukhumi nursery. Rev Roum Embryol Cytol 2:2530, 1965.

236. Volkova LS, Vyazov OE, Verbitsky MS, Lapin BA, Kooksova MI, Andreyev AV: Experimental reproduction of the hemolytic disease of the newborn in *Papio hamadryas* (Linneus). Rev Roum Embryol Cytol 3:119–130, 1966.

237. Walsh RJ, Montgomery C: A new human iso-agglutinin subdividing the MN blood groups. Nature (London) 160:504, 1947.

238. Watkins MW: Blood group substances. Science 162:172–181, 1966.

239. Weiner W, Lewis HBM, Moore P, Sanger R, Race R: A gene y modifying the blood group antigen. A Vox Sang 2:25–37, 1957.

240. Wiener AS: The agglutinogens M and N in anthropoid apes. J Immunol 34:11–18, 1938.

240a. Wiener AS: Heredity of the Rh blood types. IX. Observations in a series of 526 cases of disputed parentage. Am J Hum Genet 2:177–197, 1950.

241. Wiener AS: Blood group factors in anthropoid apes and monkeys. I. Studies on a chimpanzee "Pan." Am J Phys Anthropol 10:372–375, 1952.

242. Wiener AS: Isoagglutinins in dried blood stains. A sensitive technique for their determination. J Forensic Med 10:130–132, 1963.

243. Wiener AS: Blood groups of chimpanzee and other non human primates: their implications for the human blood groups. Trans NY Acad Sci Ser 2 27:48–50, 1965.

244. Wiener AS: New theory of antibody formation. The "prefab adaptive" theory. Lab Dig 37:5–7, 1974.

245. Wiener AS, Baldwin M, Gordon EB: Blood groups in chimpanzees. Exp Med Surg 21:159–163, 1963.

246. Wiener AS, Brancato GE: Rh hemolytic disease and auto-hemolytic disease. The possible role of agents other than antibodies in their pathogenesis. Exp Med Surg 19:344–353, 1961.

247. Wiener AS, Candela PB, Goss LJ: Blood group factors in the blood, organs and secretions of primates. J Immunol 45:229–236, 1952.

248. Wiener AS, Gavan JA, Gordon EB: Blood group factors in anthropoid apes and monkeys. II. Further studies on the Rh-Hr factors. Am J Phys Anthropol 11:39–45, 1953.

249. Wiener AS, Gordon EB: A hitherto undescribed human blood group. Am B J Haematol 2:305, 1956.

250. Wiener AS, Gordon EB: The blood groups of chimpanzees. A-B-O groups and M-N types. Am J Phys Anthropol 18:301–311, 1960.

251. Wiener AS, Gordon EB: The blood groups of chimpanzees: The Rh-Hr (CDE/cde) blood types. Am J Phys Anthropol 19:35–43, 1961.

252. Wiener AS, Gordon EB, Gallop C: Studies on autoantibodies in human sera. J Immunol 71:58–65, 1953.

253. Wiener AS, Gordon EB, Moor-Jankowski J: The Lewis blood groups in man. A review with supporting data on nonhuman primates. J Foren Med 11:67–83, 1964.

254. Wiener AS, Gordon EB, Moor-Jankowski J, Socha WW: Homologues of the human M-N blood types in gorillas and other nonhuman primates. Haematologia (Budapest) 6:419–432, 1972.

255. Wiener AS, Gordon EB, Socha W, Moor-Jankowski J: Population genetics of chimpanzee blood groups. Am J Phys Anthropol 37:301–310, 1972.

256. Wiener AS, Hurst JG, Handman L: Emploie de gelatine et d'autres produits de remplacement pour le titrage des anticorps Rh univalents par la réaction de conglutination. Rev Hématol 3:3–12, 1948.

257. Wiener AS, Moloney WC: Hemolytic transfusion reactions. IV. Differential diagnosis: "dangerous universal donor" or intragroup incompatibility. Am J Clin Pathol 13:74–80, 1943.

258. Wiener AS, Moor-Jankowski J: The V-A-B blood group system of chimpanzees: a paradox in the application of the 2 × 2 contingency test. Transfusion 5:64–70, 1965.

259. Wiener AS, Moor-Jankowski J: The A-B-O blood groups of baboons. Am J Phys Anthropol 30:117–122, 1969.

260. Wiener AS, Moor-Jankowski J: Blood groups of nonhuman primates. Cong Int Hématol, Paris, 1978.

261. Wiener AS, Moor-Jankowski J, Balner H, Gordon EB: Blood groups in monkeys, demonstrated with antisera produced by combined skin transplantation and isoimmunization with red cells. Int Arch Allergy Appl Immunol 34:386–391, 1968.

262. Wiener AS, Moor-Jankowski J, Brancato GJ: L W factor. Haematologia 3:385–393, 1969.

263. Wiener AS, Moor-Jankowski J, Cadigan FC, Gordon EB: Comparison of the ABH blood groups specificities and MN blood types in Man, gibbons *(hylobates)* and siamangs *(symphalangus)*. Transfusion 8:235–244, 1968.

264. Wiener AS, Moor-Jankowski J, Gordon EB: Blood groups of apes and monkeys. II. The A-B-O blood groups, secretor and Lewis types of apes. Am J Phys Anthropol 21:271–281, 1963.

265. Wiener AS, Moor-Jankowski J, Gordon EB: Blood group antigens and cross reacting antibodies in primates including man. I. Production of antisera for agglutination M by immunization with blood other than human M blood. J Immunol 92:391–396, 1963.

266. Wiener AS, Moor-Jankowski J, Gordon EB: Blood groups antigens and cross-reacting antibodies in primates, including man. II. Studies on the M-N types of orangutans. J Immunol 92:101–105, 1964.

267. Wiener AS, Moor-Jankowski J, Gordon EB: Three "new" blood factors of chimpanzee blood: A^c, B^c and C^c. Am J Phys Anthropol 22:413–421, 1964.

267a. Wiener AS, Moor-Jankowski J, Gordon EB: Blood groups of apes and monkeys. IV. The Rh-Hr blood types of anthropoid apes. Am J Hum Genet 16:246–253, 1964.

268. Wiener AS, Moor-Jankowski J, Gordon EB: The relationship of the H substance to the ABO blood groups. Int Arch Allergy Appl Immunol 29:82–100, 1966.

269. Wiener AS, Moor-Jankowski J, Gordon EB: Marmosets as laboratory animals. V. Blood groups of marmosets. Lab Anim Care 17:71–76, 1967.

270. Wiener AS, Moor-Jankowski J, Gordon EB: Blood groups of gorillas. Kriminal forens Wiss 6:31, 1971.

271. Wiener AS, Moor-Jankowski J, Gordon EB, Daumy OM, Davis J: Blood groups of gibbons; further observations. Int Arch Allergy Appl Immunol 30:466–469, 1966.

272. Wiener AS, Moor-Jankowski J, Gordon EB, Davis J: The blood factors I and i in primates, including man, and in lower species. Am J Phys Anthropol 23:389–396, 1965.

273. Wiener AS, Moor-Jankowski J, Gordon E, Kratochvil CL: Individual differences in chimpanzee blood, demonstrable with absorbed human anti-Rh_0 sera. Proc Natl Acad Sci USA 56:458, 1966.

274. Wiener AS, Moor-Jankowski J, Gordon EB, Riopelle AJ, Shell WF: Human type blood factors in gibbons, with special reference to the multiplicity of serological specificities of human type M blood. Transfusion 6:311–318, 1966.

275. Wiener AS, Moor-Jankowski J, Riopelle AJ, Shell WF: Simian blood groups. Another blood group system, C-E-F, in chimpanzees. Transfusion 5:508–515, 1965.

276. Wiener AS, Peters HR: Hemolytic reactions following transfusions of the homologous group with three cases in which the same agglutinogen was responsible. Ann Intern Med 13:2306–2322, 1940.

277. Wiener AS, Rosenfield RE: M^e, a blood factor common to the antigenic properties M and He. J Immunol 87:376, 1961.

278. Wiener AS, Socha WW: Macro and micro differences in blood group antigens and antibodies. II. Subgroups of A, an example of graded microdifferences. Int Arch Allergy Appl Immunol 47:946–950, 1974.

279. Wiener AS, Socha WW: Spontaneously occurring agglutinins in primate sera. II. Their classification and implications for the mechanism of antibody formation. Haematologia (Budapest) 10:463–467, 1976.

280. Wiener AS, Socha WW: Methods available for solving medico-legal problems of disputed parentage. J Forens Sci 21:42–64, 1976.

281. Wiener AS, Socha WW: "A-B-O Blood Groups and Lewis Types. Questions and Answers; Problems and Solutions. A Teaching Manual." New York: Stratton Intercontinental Medical Book Corp, 1976.

282. Wiener AS, Socha WW, Arons EB, Mortelmans J, Moor-Jankowski J: Blood groups of gorillas: further observations. J Med Primatol 5:317–332, 1976.

283. Wiener AS, Socha WW, Gordon EB: The relationship of the H substance to the A-B-O blood groups. II. Observations on Whites, Negroes and Chinese. Vox Sang 22:97–106, 1972.

284. Wiener AS, Socha WW, Gordon EB, Moor-Jankowski J: The demonstration of human Gm-like serum gamma-globulin types using non-human primate reagents. I. Experiments with anti-gibbon globulin serum. Int Arch Allergy Appl Immunol 44:140–154, 1973.

285. Wiener AS, Socha WW, Moor-Jankowski J: Homologues of the human A-B-O blood groups in apes and monkeys. Haematologia (Budapest) 18:195–216, 1974.

286. Wiener AS, Socha WW, Moor-Jankowski J: Erythroblastosis models: II. Maternofetal incompatibility in chimpanzee. Folia Primatol 27:68–74, 1977.

287. Wiener AS, Socha W, Moor-Jankowski J, Gordon EB, Kaczera Z: Blood groups of macaques. Further studies on iso-immune rhesus monkey sera, their cross-reactions and the blood factor E^{rh}. Int Arch Allergy Appl Immunol 44:140–154, 1973.

288. Wiener AS, Socha WW, Niemann W, Moor-Jankowski J: Erythroblastosis models. A review and new experimental data in monkeys. J Med Primatol 4:179–187, 1975.

289. Wiener AS, Unger LJ, Cohen L, Feldman J: Type-specific cold autoantibodies as a cause of acquired hemolytic anemia and transfusion reactions; biologic tests with bovine red cells. Ann Intern Med 44:22–240, 1956.

290. Wiener AS, Unger LG, Gordon EB: Fatal hemolytic transfusion reaction caused by sensitization to a new blood factor U. J Am Med Assoc 153:1444–1446, 1953.

291. Wiener AS, Wade M: The Rh and Hr factors in chimpanzees. Science 102:177, 1945.

292. Wiener AS, Wexler IBI, "Rh-Hr Syllabus." 2nd ed. New York: Grune and Stratton, 1963.

292a. Wiener AS, Wisecup W, Moor-Jankowski J: A "new" simian-type blood factor, L^c, associated with the C-E-F blood group system of chimpanzees. Transfusion 7:351–354, 1967.

293. Yamashita K, Tachibana Y, Takasaki S, Kobata A: ABO blood group determinant with branched cores. Nature 262:702, 1976.

293a. Yoshida A, Davé V, Branch DR, Yamaguchi H, Okubo Y: An enzyme basis for blood type A intermediate status. Am J Hum Genet 34:919–924, 1982.

294. Zmijewski CM, Metzgar RS: Production of chimpanzee isohemagglutinins by immunization with human erythrocytes. Fed Proc 23:295, 1964.

295. Zmijewski CM, Metzgar RS: Production of chimpanzee isohemagglutinins with human erythrocytes. Transfusion 5:7–11, 1965.

Index

A-B-O blood group. *See* Blood group system, A-B-O
N-Acetyl-galactosamine, 39, 40, 61
N-Acetyl-neuraminic acid, 61, 62, 63, 66
Acid citrate dextrose (ACD), 108, 143
Acid elution method, 203
Adamic theory of speciation, 228–229
Agglutination, saline, 138–139
  and spontaneously occurring isoagglutinins, 164, 167, 169
Agglutinins
  anti-A, anti-B, and anti-H, 164, 165, 167, 174–176
  cold, 176, 201
  spontaneously occurring, and serologic maternofetal incompatibility, 186, 188
  see also Heteroagglutinins, Isoagglutinins
Agglutinogen, B-like, 115, 117–119
Amerinds, 234
Anemia
  hemolytic, 179, 181, 185, 188
  sickle cell, 27, 203
Anesthesia, general, 106–107
Anthropoidea. *See* Apes, anthropoid
Antibodies
  auto-, and spontaneously occurring isoagglutinins, 172–173
  in homologous transfusion, 190–191, 197–198
  IgG and IgM, 144
  monoclonal. *See* Monoclonal antibodies
  natural, and serological maternofetal incompatibility, 186, 188
  naturally occurring, 176–177
  see also Immunoglobulins, Isoagglutinins
Anticoagulants, 108, 141
Antigens
  competition, 233–234
  intravenous administration of, in immunization, 133, 137
  low-immunogenicity, and monoclonal antibodies, 159–160
  para-, 91
  see also Blood group systems, specific antigens
Antiglobulin
  method and spontaneously occurring isoagglutinins, 164–168, 171, 176
  sera production, 138, 139, 141
  test, direct (Coombs), 181, 183, 201
Antisera
  anti-Fy, 132
  anti-LW, 98, 164, 174, 175
  anti-M, 174–175
  anti-Miltenberger, 221
  anti-P, 164, 174, 175
  anti-Rh, 175
  anti-Rh_0(D), 221
  of human origin, Rhesus system typing, 120, 125
  M-N rabbit, 102
  primate immune, 219, 221
  rhesus allo-, and transplantation, 205–206
  testing, 138–139

Antisera (*Cont'd*)
 see also Typing and typing reagents
Apes, anthropoid
 A_1 and A_2 mutations, 48
 immunization, 137
 M-N system, 55–57
 Platyrrhini cf. Catarrhini, 14
 Rhesus system chromosomes, cf.
  humans, 87–88
 taxonomy, 14–22
 see also Primates, specific species
Arboreal life, nonhuman primates, 4
Autoantibodies and spontaneously
  occurring isoagglutinins, 172–173
Aye-aye (*Daubentonia
  madagascariensis*), 13

Baboons (*Papio*)
 anti-G^p, 166
 anti-Z^p, 166
 ca and hu specificities, 155, 157
 cross-transfusion, rhesus monkey,
  189–196
 graded B^p blood group system, 94–96
  anti-B^p, 166, 167
  phylogenic connection with
   macaque D^{rh} system, 95–96
  step-like evolution, 94, 96–97
  typing reagents, 155
 hamadryas, 8
  serological maternofetal
   incompatibility, 180–183,
   186
 human anti-baboon heteroagglutinins,
  202
 spontaneously occurring
  isoagglutinins, 165–168, 173,
  174
 taxonomy, 15, 16
 typing reagents, 155, 157
Behavior, learned, 8–10
Bernard, Claude, 26
Blacks
 Aint, 48
 Hu and He antigens, 66
 Su recessive allele, 68

Blood factors. *See* Factors, blood
Blood group incompatibility in
  transplantation, 203–206; *see also*
  Serological maternofetal
  incompatibility
Blood group systems (in general)
 discovery
  in domestic animals, 35–36
  in man, 31–34
  in nonhuman primates, 36–38
 as genetic markers, 206–212
  chimpanzee, 207, 208
  see also Paternity investigation
 in genetics, importance in study, 31–32
 graded, specific to Old World
  monkeys, 91–96; *see also*
  specific species and blood group
  systems
 cf. morphological traits, 32
 nomenclature difficulties, 38
 see also specific antigens and specific
  systems
Blood group system, A-B-O
 A_m and B_m, 41–42, 44
 A, B, and H substances, 39, 40, 42,
  44, 45, 48, 233
  anti-A, anti-B, and anti-H
   agglutinins, 112, 164, 165,
   167, 174–176
  Bombay type, 40, 175
  -like, in plants and
   microorganisms, 175
  saliva inhibition test, 112, 115–120
 antigen competition, 233–234
 anti-H lectin, 42
 discovery, 31, 33–35, 37
 evolution in primates, 40–41
 four basic groups, 32
 genetic model, 39–45
  H/h pair, 39–40, 42, 44–45
  O, B, and *A*, 40–41, 44, 51
  precursor chains, 44–45
  Se gene, 42, 44–45
  successive appearance in primate
   evolution, 43
  Y and *y* genes, 41

macaques, 230
cf. M-N system, 53
mutations A_1 and A_2, 45–51, 84, 163–164, 170, 174
    A_{int} in blacks, 48
    in anthropoid apes, 48
    glycolypid types A^a-A^d, H_1-H_4, 48–51
polymorphism, variations in nonhuman primates, 51
tests
    B-like agglutinogen, 115, 117–119
    elimination of heteroagglutinins, 110, 116
    hemagglutination, 109–111, 117
    howler monkeys, 115–118, 120
    isoagglutinins anti-A and anti-B, 112
    marmosets, 115, 118, 120
    reverse grouping, 112
    saliva inhibition for A, B, and H group substances, 112, 115–120
    subgroup tests, 112
    techniques, 99–101
Blood group system, Kell, 104, 236
Blood group system, Kidd, 104
Blood group system, Lewis, 44–45
    sequential redundancies in evolution, 237
    typing, 102, 128–129
Blood group system, M-N, 34, 96, 150, 236
    cf. A-B-O, 53
    α-acetyl-galactosamine, 61
    alleles
        m lost in chimpanzee, 60, 61
        M/m and N/n, 59, 64–66
        n lost in man and gibbon, 59, 60, 63
        N lost in orangutan, 60, 64
        recessive Su, in blacks, 68
    anthropoid apes
        chimpanzee M_1^{ch} and M_2^{ch}, 55–56
        cf. human, 56–57
    antigens

Henshaw (He), 66–67, 74
Hunter (Hu), 66–67
Miltenberger series, 67–68, 70, 74
M, *Macaca mulatta*, 55
M-N series, 54–66
    M_1^{ch} and M_2^{ch}, 55–56
    S/s/U series, 68–69, 74
    V^c/A^c/B^c/D^c series (chimpanzee), 67–71
    W^c, 70–72, 74
discovery, 54–56, 75
evolution, phylogenic, of chromosome segment, 57–58
factors, 54
    A-F, 69–74
    Me, possible, 67
    M_1, 55
    N precursor of M, 58
β-galactose, 61
genetic models, 59–63, 65–66
glycophorin fragments A, B, C, 62–65
gorilla, N and MN phenotypes, 60, 62
mutations
    progressive, 65
    reserve, 58
    species-specific, 63–65
N and M receptor placement, 60–61
NANA, 61, 62, 63, 66
phylogeny, 71–74
cf. Rhesus system, 54, 75, 81
T cf. Tn polyagglutinability, 61, 62
transformation of M to N, 63–64
typing, 120, 125
Blood group system, P, 34
    discovery, 54
Blood group system, R-C-E-F,
    chimpanzee, 70, 145, 147, 150, 206, 210–212, 215
    anti-R^c (anti-L^c), 79–80
    as genetic model, 84–88, 96–98
    R^c antigen, 184, 210
    R^c-like antigen, 185
Blood group system, Rhesus, 38, 150, 236–237
    alleles, postulated, 85–86
    anti-C and anti-E, 81, 89

Blood group system (*Cont'd*)
  anti-D, 79–80
  antigens
    $C^c/c^c/E^c/F^c$, 81, 84
    D^{rh}, 191
    D (Rh_0), 77, 78, 84, 96, 98
    G, 77
  chromosomes, human cf. anthropoid
      apes, 87–88
  discovery, 55, 75–76
  evolution, 78, 89–90
  factors recognized by
    human reagents, 78–81
    simian reagents, 81, 84
  loci
    C/c, 86–87, 90
    D/d, 77
    E/e, 77, 90
    R/r, 85–88
  in man, 76–78
  cf. M-N system, 54, 75, 81
  nomenclatures, compared, 76–77
  in nonhuman primates, 78–89
    chimpanzee, R-C-E-F system,
        genetic model, 70, 84–88,
        96–98
    in gibbons, 89
    in gorilla, 88–89
    in Old World monkeys, 89
    in orangutans, 89
  precursor R, evolution, 80
  and spontaneously occurring
      isoagglutinins, 164, 165
  subgroups c_1 and c_2, 84–85
  testing, 102–103
  typing, 120, 125, 128
Blood group system, V-A-B-D,
    chimpanzee, 67–71, 105, 145, 147,
    184, 206, 210–212, 214, 236
Blood specimens
  collection, 106–108
    anticoagulants, 108
    freezing, 109
    general anesthesia, 106–107
  shipment
    coolants, 108

dry ice contraindicated, 108
    insulated containers, 109
    storage, 107–108
Blood transfusion. *See* Transfusion, blood
Bombay type, 40, 175
Breeding and serologic screening of
    nonhuman primate females, 186

Catarrhini (Old World higher primates),
    taxonomy, 15–22
  Cercopithecoidea, 15–20
  Cercopithecus (vervet or green
      monkey, guenon), 15, 18–20
  Cynopithecus, 17
  Erythrocebus, 20
  Gorilla, 21–22
  Hominoidea, 20–22
  Hylobates (gibbon), 20–21
  Macaca, 16–18
  Miopithecus, 19
  Pan (chimpanzee), 21
  Papio (baboons), 15, 16
  Pongidae, 21–22
  Pongo (orangutan), 22
  Presbytis (langurs), 20
  Symphalangus (siamang), 20, 21
  Theropithecus (geladas), 16
  see also Monkeys, specific species
Cells
  human leukemia K562, 63
  hybridoma, 159–161
Cellulose, DEAE, 139
Cercopithecoidea, taxonomy, 15, 16
Cercopithecus (vervet or green monkey,
    guenon), taxonomy, 15, 18–20
Chimpanzee *(Pan)*
  antigens
    R^c, 184, 210
    R^c-like, 185
    $V^c/A^c/B^c/D^c$ series, 67–71
  anti R^c (anti-L^c), 79–80
  blood group systems
    as genetic markers, 207, 208
    M-N, M_1^{ch} and M_2^{ch}, 55–56
    R-C-E-F, 145, 147, 150, 206, 210–
        212

R-C-E-F as genetic model, 84–88, 96–98

Rhesus, 70

Rhesus, typing, 128

V-A-B-D, 105, 145, 147, 206, 210–212, 236

see also specific blood group systems

cold agglutinins, 176

factors A-F, 69–74

glycophorin A, 64–65

homologous transfusion, 200

immunization, 141, 142

isoimmunized, in standardization of typing reagents, 145, 147

m allele lost, 60, 61

pygmy (*P. paniscus* cf. *P. troglodytes*), seroprimatology, 212–215; see also Seroprimatology

serological maternofetal incompatibility, 184

taxonomy, 21

typing reagents, 157, 158

Chromosome(s)

modifications in speciation, 229–230

segments, sequential redundancies, role in evolution, 236–237

Coefficients of heterozygosity (*H*) and polymorphism (*P*), 225, 232–233

Cold agglutinins, 176, 201; see also Isoagglutinins

Collection. See Blood specimens

Colobinae, 20

Coma, hepatic, 201, 202

Containers

insulated, 109

vacutainers, 107

Coolants, 108

Coombs (direct antiglobulin) test, 181, 183, 201

Culture, 10

Cynopithecus, taxonomy, 17

Darwin, Charles, 24, 33, 34

Daubentonia madagascariensis (aye-aye), 13

Dentition, 2, 11, 14

Development, sensorimotor, nonhuman primates, 4, 6

Dextran method, 106, 138

Direct antiglobulin test (Coombs), 181, 183, 201

Dolichos biflorus lectin (anti-A$_1$), 100–101, 112, 117, 158, 212

Domestic animals, discovery of blood groups, 35–36

Drosophila, 63, 65, 230

Dry ice contraindicated in blood shipment, 108

Duffy-related ape specificity testing, 132

Duffy technique, 104

EDTA, 108

Egg transfer experiment, 219, 220

Electrophoresis, 34–35

Environmental variety and polymorphism, 231–233, 235

Epidemics and polymorphism, 234–235

Epstein-Barr virus, 28

Erythroblastosis fetalis, 179, 180; see also Serological maternofetal incompatibility

Erythrocebus, taxonomy, 20

Evolution

A-B-O system, 40–41, 43

B^p system in baboons, 94, 96, 97

D^{rh} system in macaques, 92, 94, 96, 97

mechanisms, 223–224

M-N system, chromosome segment, 57–58

Rhesus system, 78, 89–90

precursor R, 80

role of chromosome segment sequential redundancies, 236–237

Evonymus europaeus lectin (anti-B, H), 117, 120

Exchange transfusion, 185, 186

Exons, 227–228

Factors, blood, 224–226
  A^{ba}, 157
  A-F, 69–74
  alloantigens, 225
  definition, 36
  heteroantigens, isoantigens,
      paraantigens, 224
  I and i, 103, 129, 132
  LW, 237
  M$_1$, 55
  Me, possible, 67
  monomorphic cf. polymorphic, 224,
      225
  N, precursor of M, 58
  Rh, and spontaneously occurring
      isoagglutinins, 164, 165
  Rhesus system, factors recognized by
      human reagents, 78–81
      simian reagents, 81, 84
  see also Blood group systems
Fetus. See Serological maternofetal
    incompatibility
Ficinated red cell method, 129, 138, 139,
    144, 145, 147, 157
  and Lewis typing, 129
  Rh-hr system typing, 120
  and spontaneously occurring
      isoagglutinins, 166–167, 173
Folklore, monkeys in, 23
Founder effect, 231
Freezing, blood, 109
Freund's adjuvant, 105, 133, 139, 182,
    183
L-Fucose, 39

D-Galactose, 39, 40, 61
Galen, 24
Games, nonhuman primates, 6
Geladas (Theropithecus), taxonomy, 16
Genes
  A and B, 40–41, 44, 51
  C/c, 86–87, 90
  D/d, 77
  E/e, 77, 90
  He, 66–67
  H/h pair, 39–40, 42, 44–45

Hu, 66–67
M allele lost in chimpanzee, 60, 61
M/m, 59, 64–65
n allele lost in man and gibbon, 59,
    60, 63
N allele lost in orangutan, 60, 64
N/n, 59, 64–65
O, 40–41, 44, 51
postulated alleles of Rhesus system,
    85–86
R/r, 85–88
Se, 42, 44–45
Su, recessive allele in blacks, 68
Y and y, 41
see also under specific blood group
    systems
Genetic drift, 230–231
Genetic markers, blood groups as, 206–
    212
  chimpanzee, 207, 208
  definition, 206–207
  see also Paternity investigation
Genetic polymorphism, advantages, 35–
    36; see also Polymorphism
Genetics, importance of blood groups in
    study, 31–32; see also
    Seroprimatology
Gibbon (Hylobates), 7
  n allele lost, 59, 60, 63
  Rhesus system, 89
  taxonomy, 20–21
  typing reagents, 157
Glycolipids types A^a-A^d, H$_1$-H$_4$ in A-B-O
    blood group mutations, 48–51
Glycophorin A, chimpanzee, 64–65
  fragments A, B, C, 62–64
Gorilla
  chimpanzee V^c/A^c/B^c/D^c series
      antigens in, 69
  N and MN phenotypes, 60, 62
  Rhesus system, 89
  typing, 128
  taxonomy, 21–22
Green monkey (Cercopithecus), 14, 18–20
Guenon (Cercopithecus), taxonomy, 15,
    18–20

Hamadryas baboons. *See* Baboons *(Papio)*
Hands, nonhuman primates, 2, 4, 6
Harem structure, nonhuman primates, 8
Helix pomatia (snail) lectin (anti-A), 101, 117
Hemagglutination
  and A-B-O blood group tests, 109-111, 117
  microtiter, in antisera testing, 138
Hematocrit and transplantation, 203
Hemolytic anemia, 179, 181, 185, 188
Henshaw (He) antigens, 66-67, 74
Hepatic coma, 201, 202
Herpes viruses, 28
Heteroagglutinins, 99, 100, 163, 164
  elimination, and A-B-O blood group tests, 110, 116
  human anti-baboon, 202
  and Rhesus system typing, 125
Heterohemolysins, 163, 164
Heterosis, 36
Heterostasis, 59
Heterozygosity, coefficient of *(H)*, 225, 232-233
Histocompatibility (HLA), 38, 205, 236
Hoffstetter classification of primates, 22
Hominization, 235, 236
Hominoidea, 2
  taxonomy, 20-22
Homologous transfusion. *See* Transfusion, blood
Homo sapiens. See Man
Hopeful monster, 228
Howler monkeys, A-B-O blood group tests, 15, 115-118, 120; *see also* Platyrrhini
Human. *See* Man
Hunter (Hu) antigens, 66-67
Hybridoma cells, 159-161
Hybrids, luxuriance of, 36
Hylobates. See Gibbon *(Hylobates)*
Hyperbilirubinemia, 181, 184, 185

Icterus gravis, 179, 181
Immunization of primates, 133, 137
  anthropoid apes, 137

antiglobulin sera production, 138, 139, 141
antisera testing, 138-139
  controls, 138
  methods, 138, 139
  microtiter hemagglutination, 138
Freund's complete and incomplete adjuvants, 133, 139
injection of antigen
  intramuscular, 133, 139
  intraperitoneal, 137
  intravenous, 133, 137
protocol, 141-143
tuberculin tests, 133
Immunoglobulins
  G, 139, 176, 184, 186, 190
  IgG and IgM antibodies, 144
  M, 174, 175
Immunohematology, 225
  cf. paleontology, 224
"Immunological perspective," 100, 225-226
Immunological polymorphism, 32-35
  M-N group, 64
Immunology, 27-28
Inborn errors of metabolism, 207
Inbreeding in nonhuman primates, 206
Incompatibility, serological maternofetal. *See* Serological maternofetal incompatibility
Insects, cf. primates, 8-9
Intelligence, nonhuman primates, 6
Introns, 227-228
Isoagglutinins, anti-A and anti-B, 112
Isoagglutinins, spontaneously occurring, 163-177
  antiglobulin method, 164-168, 171, 176
  autoantibodies, 172-173
  classification and significance, 173-177
  cold, 164, 173, 174
  cold and room but not body temperature, 174
  ficinated red cell method, 166-167, 173

Isoagglutinins (*Cont'd*)
  irregular, 163–164
  reactive at room temperature, 175–177
  Rh factor, 164, 165
  saline agglutination, 164, 167, 169
  in various species, 165–173
    polymorphism, 168, 172
Isoimmunization, difficulty of, 105–106
Isolation, sexual, and speciation, 229

Kell system, 236
Kell technique, 104
Kernicterus, 179, 181
Ketamine hydrochloride (Vetalan), 107
Kidd technique, 104
Kidney transplantation, 203
Kuru, 28

Landsteiner, K., 31, 33, 54, 75, 163–164,
    225
Landsteiner's rule, 101, 112
Langurs (*Presbytis*)
  entellus, 9
  taxonomy, 20
Learned behavior, 8–10
Lectin(s)
  anti-H, 42
  Dolichos biflorus (anti-A₁), 100–101,
    112, 117, 158, 212
  Evonymus europaeus (anti-B, H), 117,
    120
  lima bean *(Phaseolus vulgarus)* (anti-
    A), 100, 117
  snail *(Helix pomatia)* (anti-A), 101,
    117
  tests, 100–101
  Ulex europaeus (anti-H), 101, 115,
    117, 158
  Vicia gramina and *V. unijuga*, 56, 58,
    61, 64, 102, 120, 158
Lemuriformes, 11–13
Leukemia, 29
  human, K562 cells, 63
Lewis. *See* Blood group system, Lewis
Lima bean lectin *(Phaseolus vulgarus)*
  (anti-A), 100, 117

Limbs, nonhuman primates, 2, 4, 6
Linnaeus, 24, 33
Lorisiformes, 11, 13
Low ionic strength solutions (LISS), 106
Luxuriance of hybrids, 36
Lymphocytes, T, 161
Lymphomatous disease, 176

Macaques *(Macaca)*
  A-B-O system, 230
  egg transfer experiment, 219–220
  graded D^rh blood group system, 91–
    94, 219
    as genetic marker, in pig-tailed
      macaques, 208–209
    genetics in *M. nemestrina*, 95
    step-like evolution, 92, 94, 96, 97
    suggested nomenclature, 92
    typing reagents, 92
  habitat, 16–17
  Japanese, 9
  M antigens, 55
  pig-tailed, A-B-O and D^rh blood
      groups as genetic markers, 208–
      209
  rhesus monkey *(M. mulatta)*
    alloantisera and transplantation,
      205–206
    immunization, 141, 142
    isoimmune sera, 153–155
  Rhesus system, discovery, 55
  serological maternofetal
      incompatibility, 182–183, 185–
      186
  seroprimatology, 215, 219
  spontaneously occurring
      isoagglutinins, 165, 167–173
  taxonomy, 17–18
  typing reagents, 153–155
Madagascar, 13
Man
  discovery of blood groups, 31–34
  M-N system, 96
    cf. anthropoid apes, 56–57
  n allele lost, 59, 60, 63
  Rhesus system, 76–78

chromosomes, cf. anthropoid apes, 87–88
see also Primates, specific blood group systems
Marmosets, 14
A-B-O blood group tests, 115, 118, 120
Maternofetal incompatibility. *See* Serological maternofetal incompatibility
Metabolic disorders, 29
Metabolism, inborn errors of, 207
Method
acid elution, 203
antiglobulin, and spontaneously occurring isoagglutinins, 164–168, 171, 176
dextran, 106, 138
ficinated red cell. *See* Ficinated red cell method
Miltenberger series antigens, 67–68, 70, 74
anti-Miltenberger primate immune sera, 221
Miopithecus, taxonomy, 19
MN blood group. *See* Blood group system, M-N
Monkeys
as caricature of human, 23
in folklore, 23
in history, 24–25
howler, A-B-O blood group tests, 115–118, 120
New World (Cebidae), taxonomy, 14–15
Old World (Catarrhini)
A-B-O blood group system, cf. Platyrrhini, 40–41
graded blood systems specific to, 91–96
relationship to anthropoid apes and man, 98
Rhesus system, 89
taxonomy, 15–22
patas, 20
rhesus

cross-transfusion, baboon, 189–196
immunization, 141, 142
vervet or green monkey *(Cercopithecus)*, taxonomy, 15, 18–20
worship of, 23
see also Primates, specific species
Monoclonal antibodies, 158–161
cross-reactivity, 160, 161
low-immunogenicity antigens, 159–160
methodology, 159
prohibitive cost and specificity, 160
in study of phylogenetic relationships, 161
see also Antibodies
Morphology, cf. blood groups, 32
Muller, Fritz, 51
Mutationists, 228
Muzzles, nonhuman primates, 4

Neosequences, 54, 74, 227–228, 237
Niche, 231, 235
Nomenclature, blood group, 38
macaque D^{rh} system, 92
Rhesus system, 76–77

Orangutan *(Pongo)*
N gene lost, 60, 64
Rhesus system, 89
typing, 128
serological maternofetal incompatibility, 184–185
taxonomy, 22

Paleontology, 224
Paleosequences, 54, 74, 227–228, 236–237
Pan. See Chimpanzee *(Pan)*
Papio. See Baboons *(Papio)*
Para-antigens, 91
Parasites, 27–28
Patas monkeys, 20
Paternity investigation, blood group genetic markers, 206, 208–212
A-B-O and D^{rh}, pig-tailed macaques, 208–209

P blood group. *See* Blood group system,
  P
Phaseolus vulgarus (lima bean) lectin,
  100, 117
Phenycyclidine hydrochloride (Sernylan),
  107
Phylogenic distance, 225, 226
Plants and microorganisms, A-, B-, and
  H-like antigens, 175
Plasmapheresis, 143
Platyrrhini, 14–15
  A-B-O blood group system, cf.
    Catarrhini, 40–41
  Callitrichidae, 14
  Cebidae (New World monkeys), 14–15
  see also Monkeys, specific species
Polymorphism
  causes of maintenance, 230–235
  coefficient of *(P)*, 225, 232–233
  and environmental variety, 231–233,
    235
  and epidemics, 234–235
  genetic
    advantages, 35–36
    and speciation, 228–230
  immunological, 32–35
    M-N group, 64
  interspecific cf. intraspecific, 225
  measurements in various nonhuman
    primates, 232
  neutralism, 235
  and social structure, 233–235
  spontaneously occurring
    isoagglutinins, 168, 172
Polyvinyl pyrrolidine, 138
Pongidae, taxonomy, 21–22
Pongo. See Orangutan *(Pongo)*
Predators of nonhuman primates, 7
Presbytis (langurs)
  entellus, 9
  taxonomy, 20
Primates
  arboreal life, 4
  blood groups. *See* Blood group
    systems
  dentition, 2
  evolution of A-B-O system, 40–41, 43

games, 6
immunization. *See* Immunization of
  primates
inbreeding, 206
cf. insects, 8–9
intelligence, 6
karyotypes, 25
learned behavior, importance, 8–10
limbs and hands, 2, 4, 6
morphologic and cellular
  resemblances, 25
muzzles, 2, 6
origins, 1–2, 11
predators, 7
reproductive pattern, 7
sensorimotor development, 4, 6
social structure, 6–9
  harem, 8
  and polymorphism, 233–235
specialization, lack of, 2–4
taxonomy, 11–22
  Anthropoidea, 14–22
  classification (Hoffstetter), 22
  Prosimii, 11–14
  table, 12
upright position, 6
use in medicine, 26–29
see also Apes, anthropoid; Monkeys;
  specific species
Primatology. *See* Seroprimatology
Prosimians, 1, 7
Prosimii, taxonomy, 11–14
Protoculture, 9
Purgatorius, 1

Rabbits
  immunization, 100
  M-N antisera, 102
Reagents, typing. *See* Typing and typing
  reagents
Receptor placement, N and M, 60–61
Red cell survival
  after heterologous transfusion, 202–
    203
  after homologous transfusion, 196–199
  see also Ficinated red cell method

Reproductive pattern, nonhuman
     primates, 7
Rhesus monkey. *See under* Macaques
     (Macaca)
Rhesus system. *See* Blood group system,
     Rhesus

Saline agglutination, 138–139
     and spontaneously occurring
          isoagglutinins, 164, 167, 169
Saliva
     inhibition tests, 102, 128–129
          for A, B, and H group substances,
               112, 115–120
     samples, 108
Screening, serologic, of nonhuman
     primate females before breeding,
     186
Sensorimotor development, nonhuman
     primates, 4, 6
Sephadex G-25, 139
Sequences, paleo- and neo-, 54, 74, 227–
     228, 236–237
Sera, primate immune, 219, 221
Sernylan (phenycyclidine hydrochloride),
     107
Serological maternofetal incompatibility,
     179–188
     chimpanzees, 184
     in hamadryas baboons, 180–183, 186
     in vivo survival of incompatible red
          cells, 183–184, 196–197
     in macaques, 182–183, 185–186
     natural antibodies, 185–186
     orangutans, 184–185
     protective mechanisms, 180, 183
     spontaneously occurring agglutinins,
          186, 188
Seroprimatology, 212–219
     macaques, 215, 219
          D^{rh} series, 219
          egg transfer experiment, 219, 220
     populations, comparative study of
          blood groups, 223–224
     pygmy chimpanzees, 212–215
          MN system, 214
          R-C-E-F system, 215

subgroup ^{A}int, 212, 214
V-A-B-D blood group distribution,
     214
taxonomic identification, 212
Sexual isolation and speciation, 229
Shipment. *See under* Blood specimens
Shrews, tree, 13
Siamang *(Symphalangus)*, taxonomy, 20,
     21
Sickle cell anemia, 27, 203
Smell, nonhuman primates, 4
Snail *(Helix pomatia)* lectin (anti-A), 101,
     117
Social structure
     nonhuman primates, 6–9
     and polymorphism, 233–235
Solutions, low ionic strength (LISS), 106
Specialization, lack of, nonhuman
     primates, 2–4
Speciation
     Adamic theory, 228–229
     chromosome modifications, 229–230
     and genetic polymorphism, 228–230
     and sexual isolation, 229
Standardization. *See under* Typing and
     typing reagents
Storage, blood specimens, 107–108
Symphalangus (siamang), taxonomy, 20,
     21

Tarsiiformes, 11, 13–14
Taxonomy
     Anthropoidea, 14–22
     Prosimii, 11–14
Temperature and spontaneously occurring
     isoagglutinins, 164, 173–177
Test(s)
     A-B-O system. *See under* Blood group
          system, A-B-O
     difficulty of isoimmunization, 105–106
     direct antiglobulin (Coombs), 181,
          183, 201
     Duffy, 104, 132
     factors I and i, 103, 129, 132
     heteroagglutinins, 99, 100
     immunization of rabbits, 100
     Kell, 104

Test(s) (*Cont'd*)
  Kidd, 104
  lectins, 100–101
  M-N rabbit antisera, 102
  reagents produced by primate
    immunization, 104–106
  Rh testing, 102–103
  saliva inhibition, 102, 128–129
  see also specific techniques, reagents,
    blood group systems
Theropithecus (geladas), taxonomy, 16
T lymphocytes, 161
Transfusion, blood, 33–34
  exchange, 185, 186
  heterologous, 201–203
    A-B-O incompatibility, 201
    cross-circulation in human hepatic
      coma, 201, 202
    human anti-baboon
      heteroagglutinins, 202
    survival of human red cells in
      primate circulation, 202–203
  homologous, 188–201
    A-B-O incompatibility, 200
    antibodies, pre-existing, 197
    antibody production, 190–191, 197–198
    antigens, D^{rh}, 191
    cardiovascular surgery, 188
    changes in hematologic values, 196
    chimpanzee, 200
    cross-match testing, 200–201
    cross-transfusion, rhesus monkey-
      baboon, 189–196
    donor preselection, 200
    in nonhuman primate
      experimentation, 188, 198
    red cell specificities,
      immunogenicity, 200
    survival of red cells, 196–199
    timing, 191
Transplantation and blood group
  compatibility, 203–206
Trees, and nonhuman primates, 4
Tree shrews, 13
Tuberculin tests, 133
Typing and typing reagents, 153–158
  anti-A, anti-B, 219, 221

baboons, 155, 157
B^P graded blood group system, 155
chimpanzees, 157, 158
factor A^{ba}, 157
gibbons, 157
Lewis, 128–129
macaques, 92, 153–155
M-N system, 120, 125
  batteries, 120
originally for typing human blood, 109
primate immune sera as, 219, 221
produced by primate immunization,
  104–106, 132
rhesus isoimmune sera, 153–155
Rhesus system, 120, 125, 128
  human reagents, 78–81
  simian reagents, 81, 84
standardization, 143–152
  anti-O^c and anti-P^c, 147
  complement interference, 150–151
  cross-immunization, 151
  isoimmunized chimpanzee, 145, 147
  macaque isoimmune serum, 151
  serologic criterion of unity, 151
  titration, 143–145
  see also specific blood group
    systems
Tyson, 24

Ulex europaeus (anti-H) lectin, 105, 115,
  117, 158
Upright position, nonhuman primates, 6

Vacutainers, 107
Vervet monkey *(Ceropithecus)*, 15, 18–20
Vesalius, 24
Vetalan (ketamine hydrochloride), 107
Vicia graminea and *V. unijuga* lectins, 56,
  58, 61, 64, 69, 102, 120, 158
Virology, 26–28
Virus
  Epstein-Barr, 28
  herpes, 28
Vision, nonhuman primates, 4

Wiener model of M-N system, 59–63
Worship of monkeys, 23